The transport of macromolecules in membrane-bounded vesicles is a fundamental process of all eukaryotic cells. The paths taken by this vesicular traffic have been intensively researched in animal cells but are less well characterised in plant cells. Nevertheless, with the development and combination of techniques in biochemistry, molecular biology and cell biology, progress in the study of plant vesicle traffic has been achieved in recent years. This book highlights the recent advances made and explores avenues for future research.

SOCIETY FOR EXPERIMENTAL BIOLOGY
SEMINAR SERIES: *45*

ENDOCYTOSIS, EXOCYTOSIS AND VESICLE TRAFFIC IN PLANTS

SOCIETY FOR EXPERIMENTAL BIOLOGY SEMINAR SERIES

A series of multi-author volumes developed from seminars held by the Society for Experimental Biology. Each volume serves not only as an introductory review of a specific topic, but also introduces the reader to experimental evidence to support the theories and principles discussed, and points the way to new research.

2. Effects of pollutants on aquatic organisms. *Edited by A. P. M. Lockwood*
6. Neurones without impulses: their significance for vertebrate and invertebrate systems. *Edited by A. Roberts and B. M. H. Bush*
8. Stomatal physiology. *Edited by P. G. Jarvis and T. A. Mansfield*
10. The cell cycle. *Edited by P. C. L. John*
11. Effects of disease on the physiology of the growing plant. *Edited P. G. Ayres*
12. Biology of the chemotactic response. *Edited by J. M. Lackie and P. C. Williamson*
14. Biological timekeeping. *Edited by J. Brady*
15. The nucleolus. *Edited by E. G. Jordan and C. A. Cullis*
16. Gills. *Edited by D. F. Houlihan, J. C. Rankin and T. J. Shuttleworth*
17. Cellular acclimatisation to environmental change. *Edited by A. R. Cossins and P. Sheterline*
19. Storage carbohydrates in vascular plants. *Edited by D. H. Lewis*
20. The physiology and biochemistry of plant respiration. *Edited by J. M. Palmer*
21. Chloroplast biogenesis. *Edited by R. J. Ellis*
23. The biosynthesis and metabolism of plant hormones. *Edited by A. Crozier and J. R. Hillman*
24. Coordination of motor behaviour. *Edited by B. M. H. Bush and F. Clarac*
25. Cell ageing and cell death. *Edited by I. Davies and D. C. Sigee*
26. The cell division cycle in plants. *Edited by J. A. Bryant and D. Francis*
27. Control of leaf growth. *Edited by N. R. Baker, W. J. Davies and C. Ong*
28. Biochemistry of plant cell walls. *Edited by C. T. Brett and J. R. Hillman*
29. Immunology in plant science. *Edited by T. L. Wang*
30. Root development and function. *Edited by P. J. Gregory, J. V. Lake and D. A. Rose*
31. Plant canopies: their growth, form and function. *Edited by G. Russell, B. Marshall and P. G. Jarvis*
32. Developmental mutants in higher plants. *Edited by H. Thomas and D. Grierson*
33. Neurohormones in invertebrates. *Edited by M. Thorndyke and G. Goldsworthy*
34. Acid toxicity and aquatic animals. *Edited by R. Morris, E. W. Taylor, D. J. A. Brown and J. A. Brown*
35. Division and segregation of organelles. *Edited by S. A. Boffey and D. Lloyd*
36. Biomechanics in evolution. *Edited by J. M. V. Rayner and R. J. Wootton*
37. Techniques in comparative evolution: An experimental approach. *Edited by C. R. Bridges and P. J. Butler*
38. Herbicides and plant metabolism. *Edited by A. D. Dodge*
39. Plants under stress. *Edited by H. G. Jones, T. J. Flowers and M. B. Jones*
40. *In situ* hybridisation: application to developmental biology and medicine. *Edited by N. Harris and D. G. Wilkinson*
41. Physiological strategies for gas exchange and metabolism. *Edited by A. J. Woakes, M. K. Grieshaber and C. R. Bridges*
42. Compartmentation of plant metabolism in non-photosynthetic tissues. *Edited by M. J. Emes*
43. Plant Growth: interactions with nutrition and environment. *Edited by J. R. Porter and D. W. Lawlor*
44. Feeding and the texture of foods. *Edited by J. F. V. Vincent and P. J. Lillford*
45. Endocytosis, exocytosis and vesicle traffic in plants. *Edited by C. R. Hawes, J. O. D. Coleman and D. E. Evans*

ENDOCYTOSIS, EXOCYTOSIS AND VESICLE TRAFFIC IN PLANTS

Edited by

C.R. Hawes

School of Biological and Molecular Sciences, Oxford Polytechnic

J.O.D. Coleman and D.E. Evans

Department of Plant Sciences, University of Oxford

CAMBRIDGE UNIVERSITY PRESS

Cambridge

New York   Port Chester

Melbourne   Sydney

Published by the Press Syndicate of the University of Cambridge
The Pitt Building, Trumpington Street, Cambridge CB2 1RP
40 West 20th Street, New York, NY 10011-4211, USA
10 Stamford Road, Oakleigh, Melbourne 3166, Australia

First published 1991

Printed in Great Britain at the University Press, Cambridge

British Library cataloguing in publication data
Endocytosis, exocytosis and vesicle traffic in plants.
1. Plants. Cells (Organisms)
I. Hawes, C. R. (Chris R.) II. Coleman, J. O. D. III. Evans, D. E.
IV. Series
581.874

Library of Congress cataloguing in publication data
Endocytosis, exocytosis and vesicle traffic in plants / edited by C. R. Hawes, J. O. D. Coleman, and D. E. Evans.
p. cm. – (Seminar series / Society for Experimental Biology ; 45)
ISBN 0-521-32844-6
1. Plant cells and tissues – Inclusions. 2. Plant translocation. 3. Endocytosis. 4. Exocytosis. 5. Coated vesicles. I. Hawes, C. R. II. Coleman, J. O. D. (Julian O. D.) III. Evans, D. E. (David E.) IV. Title: Vesicle traffic in plants. V. Series: Seminar series (Society for Experimental Biology (Great Britain)); 45.
QK725.E53 1991
574.87′4 – dc20 91-7436 CIP

ISBN 0 521 38108 8 hardback

CONTENTS

Contributors

K. BALUSEK
Abteilung Cytologie, Pflanzenphsiologisches Institut, Universität Göttingen, Untere Karspüle 2, D-3400 Göttingen, Germany
D. S. BUSH
Department of Plant Biology, University of California, Berkeley, CA 94720, USA
D. CLARKSON
AFRC Institute for Arable Crops Research, University of Bristol, Long Ashton Research Station, Bristol BS18 4AF
L. COLE
School of Biological and Molecular Sciences, Oxford Polytechnic, Gipsy Lane, Headington, Oxford OX3 0BP, UK
J. O. D. COLEMAN
Department of Plant Sciences, University of Oxford, South Parks Road, Oxford OX1 3RA, UK
H. DEPTA
Abteilung Cytologie, Pflanzenphsiologisches Institut, Universität Göttingen, Untere Karspüle 2, D-3400 Göttingen, Germany
D. E. EVANS
Department of Plant Sciences, University of Oxford, South Parks Road, Oxford OX1 3RA, UK
L. C. FOWKE
Department of Biology, University of Saskatchewan, Saskatoon, Canada S7N OWO
M. E. GALWAY
Department of Biology, University of Saskatchewan, Saskatoon, Canada S7N OWO
T. H. GIDDINGS
Department of Molecular, Cellular and Developmental Biology, University of Colorado, Boulder, CO 80309-0347, USA

N. HARRIS
Department of Biological Sciences, University of Durham, Durham DH1 3LE, UK
C. R. HAWES
School of Biological and Molecular Sciences, Oxford Polytechnic, Gipsy Lane, Headington, Oxford OX3 0BP, UK
B. HOH
Abteilung Cytologie, Pflanzenphsiologisches Institut, Universität Göttingen, Untere Karspüle 2, D-3400 Göttingen, Germany
S. E. H. HOLSTEIN
Abteilung Cytologie, Pflanzenphsiologisches Institut, Universität Göttingen, Untere Karspüle 2, D-3400 Göttingen, Germany
D. HORSLEY
Department of Plant Sciences, University of Oxford, South Parks Road, Oxford OX1 3RA, UK
R. L. JONES
Department of Plant Biology, University of California, Berkeley, CA 94720, USA
K. KRONESTEDT-ROBARDS
Institute for Applied Biology, University of York, York YO1 5DD, UK
S. LEVY
Department of Molecular, Cellular and Developmental Biology, University of Colorado, Boulder, CO 80309-0347, USA
M. A. LYNCH
Department of Molecular, Cellular and Developmental Biology, University of Colorado, Boulder, CO 80309-0347, USA
P. J. MOORE
Department of Molecular, Cellular and Developmental Biology, University of Colorado, Boulder, CO 80309-0347, USA
D. O'DRISCOLL
Department of Botany, University College, Belfield, Dublin 4, Ireland
K. OPARKA
Scottish Crop Research Institute, Invergowrie, Dundee DD2 5DA, UK
A. W. ROBARDS
Institute for Applied Biology, University of York, York YO1 5DD, UK
D. G. ROBINSON
Abteilung Cytologie, Pflanzenphsiologisches Institut, Universität Göttingen, Untere Karspüle 2, D-3400 Göttingen, Germany
L. A. STAEHELIN
Department of Molecular, Cellular and Developmental Biology, University of Colorado, Boulder, CO 80309-0347, USA

M. W. STEER
Department of Botany, University College, Belfield, Dublin 4, Ireland
P. L. STEPONKUS
Department of Agronomy, 609 Bradfield Hall, Cornell University, Ithaca, NY 14853, USA
L. STICHER
Department of Plant Biology, University of California, Berkeley, CA 94720, USA
K. M. M. SWORDS
Department of Molecular, Cellular and Developmental Biology, University of Colorado, Boulder, CO 80309-0347, USA
M. A. TANCHAK
Department of Biochemistry, University of Ottawa, Ottawa, Canada K1N 6N5
M. D. WATSON
Department of Biological Sciences, University of Durham, Durham DH1 3LE, UK
K. M. WRIGHT
Scottish Crop Research Institute, Invergowrie, Dundee DD2 5DA, UK

Preface

Eukaryotic cells need to transport macromolecules from sites of synthesis to sites of action and from sites of action to sites of degradation. Many of these molecules are translocated in vesicles. This vesicle traffic requires recognition and sorting mechanisms as well as guidance systems to ensure that vesicles and their molecular cargoes arrive at the right destinations at the right time. The paths taken by this vesicular traffic have been intensively researched in animal cells but are less well characterised in plant cells. Nevertheless in recent years important progress has been made in some areas, particularly in the study of secretory and endocytotic pathways.

This volume is an attempt to highlight current progress in the study of plant vesicle traffic, and to explore avenues for future research. The chapters in this book were based on papers presented at a symposium held during the Society for Experimental Biology Annual Meeting at the University of Warwick, April 1990.

Progress in this field of research has been made possible by combining the techniques of biochemistry, molecular biology and cell biology and is reflected in the chapters of this volume. Authors were chosen for their research expertise in the vesicular transport activities of plant cells and were given the opportunity to present their work in their own style and the freedom to air their own views.

We thank all the contributors for making our editorial task easy by submitting carefully prepared manuscripts of high quality. We hope that this volume will be of use to researchers working both on plant and on animal cellular biology and will stimulate the interest of advanced undergraduates reading plant biology.

Finally, we are indebted to all those who have contributed to this volume, directly and indirectly, and especially to the authors, the Society for Experimental Biology and to the DuPont company for their sponsorship of the meeting.

Chris Hawes, David Evans and Julian Coleman
Oxford, August 1990

C.R. HAWES, D.E. EVANS
AND J.O.D. COLEMAN

An introduction to vesicle traffic in eukaryotic cells

The multiplicity of membranes and membrane-bounded compartments of eukaryotic cells is a feature which distinguishes them from prokaryotic cells. Each compartment has a characteristic macromolecular make up and performs specific functions. Such compartmentalisation carries with it certain functional advantages but requires sophisticated mechanisms of integration.

In the case of the endomembrane system, where synthesised macromolecules are sequentially processed and packaged for final distribution either to the plasma membrane or to the lysosomal/vacuolar compartments, communication is via membrane-bounded vesicles. In a similar fashion some macromolecules entering the cell do so in the protective environment of an endocytotic vesicle. Within the endomembrane system (comprising the endoplasmic reticulum (ER), Golgi apparatus, lysosome/vacuole and endosome), the cell has evolved a sophisticated series of mechanisms for synthesising, modifying, sorting and targeting biosynthetic products to the right place at the right time. Concurrently the membrane economy of the cell is in part regulated by turnover of the plasma membrane by endocytotic vesicles, whilst signal and nutritional molecules can be internalised within the same system of vesicles.

Over the past decade major advances have been made at the cell and molecular levels in expanding our understanding of the various mechanisms involved in the synthesis, transport and targeting of membrane-bounded macromolecules and in the receptor-mediated internalisation of physiologically important molecules. However, it is still not clear whether the problems of vesicle trafficking and targeting are solved by the same mechanisms in all eukaryotic cells or whether different cell types have evolved mechanisms particular to the nature of their secretory/endocytotic molecules.

In this introductory chapter we review the recent significant advances that have been made in the study of vesicle traffic in eukaryotic cells, with an emphasis on animal cells. At the same time we indicate briefly the areas in which advances have recently been made in the study of vesicles and transport in plants. These are dealt with in greater depth in the various chapters of this volume.

Transport to the Golgi

Proteins for secretion and lysosomal proteins are synthesised on the rough ER and are co-translationally translocated to the lumen of the ER, directed by a transmembrane signal peptide.

In animal cells, transport of newly processed protein from the ER to the Golgi is via a non-signalled pathway which is mediated by transition vesicles. In spite of this lack of signal the resident soluble enzymes in the ER are selectively retained within the lumen of the organelle. This selective retention is maintained by a specific retrograde transport pathway from the Golgi back to the ER (Pelham 1988). By this means the ER can salvage, from the Golgi, soluble enzymes which are not destined for transport but which are involved in the processing of secretory proteins (e.g. protein disulphide isomerase and heavy-chain binding proteins 'BiP'; Munro & Pelham 1987). These processing proteins have the highly conserved signal sequence KDEL (Lys-Asp-Glu-Leu) at their carboxy termini, which target them back into the ER (Munro & Pelham 1987). A membrane protein of 72 000 M_r has been identified by use of anti-idiotype antibodies raised against carboxy terminal KDEL sequence monoclonals (Vaux *et al.* 1990) and this protein has the necessary characteristics to act as the KDEL receptor. A similar system is in operation in yeast cells which use HDEL (Hist-Asp-Glu-Leu) rather than KDEL as the retention signal (Pelham *et al.* 1988; Dean & Pelham 1990) and a gene, *ERD2*, encoding the HDEL receptor has been characterised (Semenza *et al.* 1990). Lippincott-Schwartz *et al.* (1990) have demonstrated the presence of an intermediate salvage compartment, which is marked by an antigen of 53 000 M_r, between the rough endoplasmic reticulum and the *cis*-Golgi and it can be assumed that it is within this compartment that sorting of ER and secretory proteins occurs (Fig. 1). It was also suggested that this retrograde transport back to the rough endoplasmic reticulum is mediated by microtubules (Lippincott-Schwartz *et al.* 1990).

In comparison to the situation in animal and yeast cells, information on ER–Golgi transport in plants is scant (see Fig. 2). In systems which predominantly secrete structural carbohydrates (e.g. wall material), perhaps no direct continuity or massive flow of biosynthetic products from the ER to the Golgi is required. Certainly, dictyosomes can exist as discrete entities, with no evidence of transition vesicles (Juniper *et al.* 1982). However, this still leaves unresolved the problem of membrane

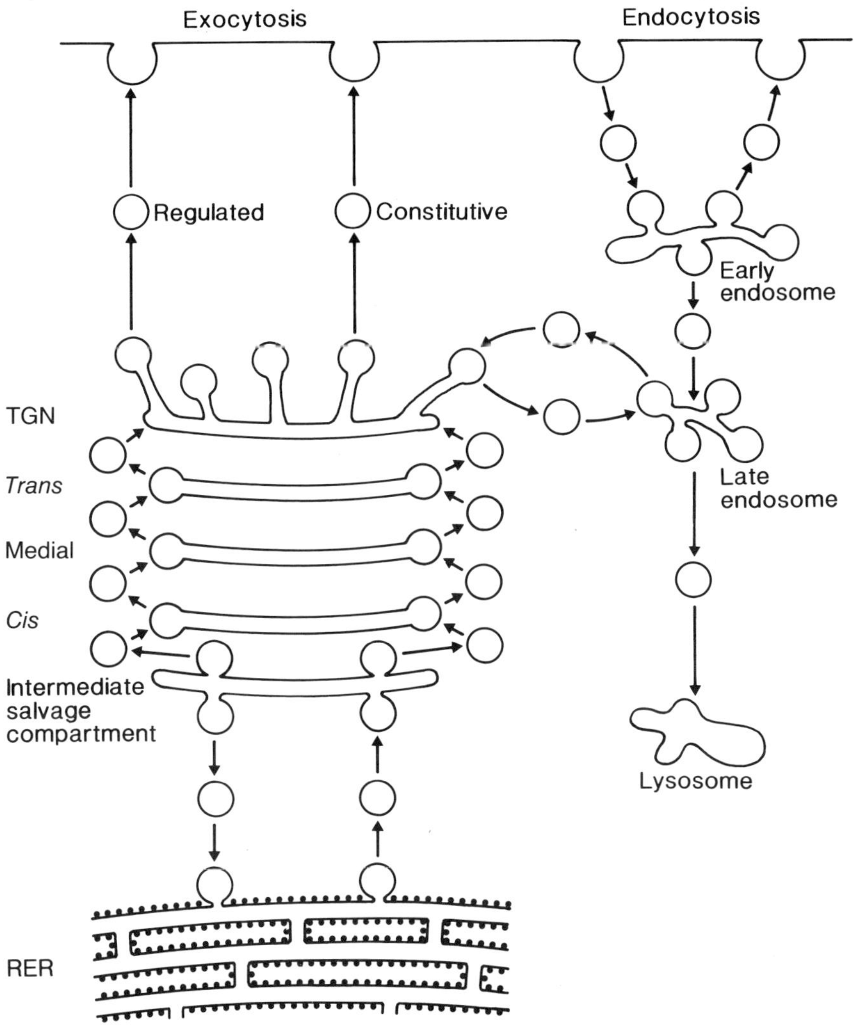

Fig. 1. Pathways for the vesicular transport of macromolecules in animal cells. TGN, *trans*-Golgi network; RER, rough endoplasmic reticulum.

flow from the synthetic sites on the ER to the Golgi. In protein-transporting systems such as developing seed storage vacuoles it is assumed that transition vesicles transport proteins from ER to the Golgi. All the biochemical evidence points to this, although good visualisation and isolation of such transport vesicles is yet to be achieved. It has also been suggested that, in germinating seeds, direct interconnections exist

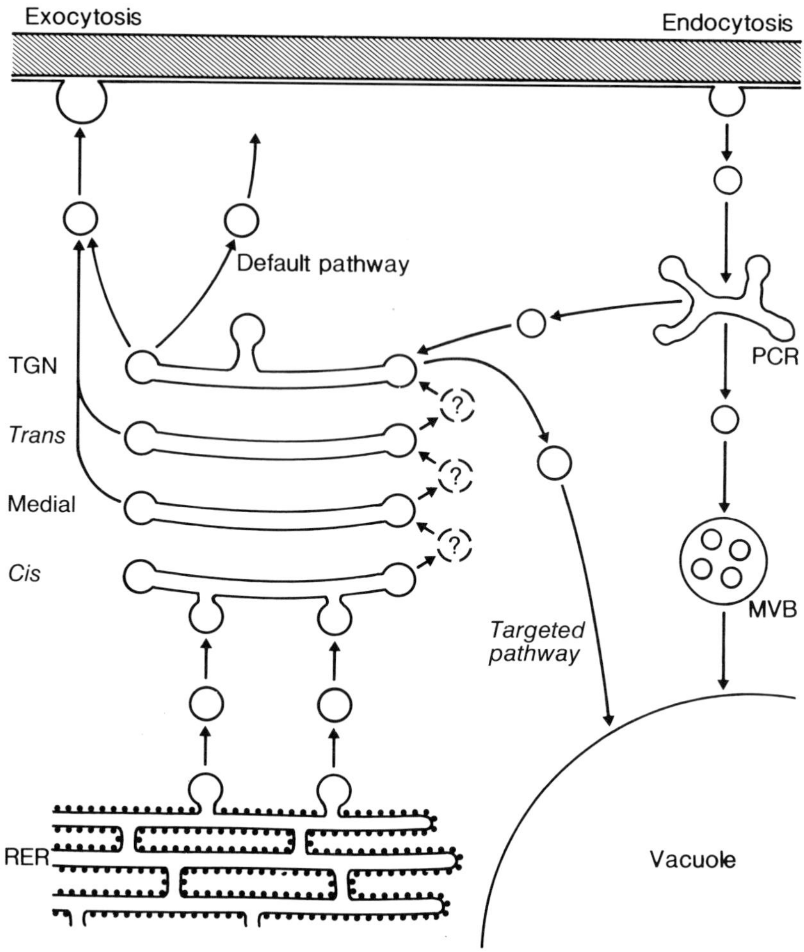

Fig. 2. Possible pathways for the vesicular transport of biosynthetic proteins and carbohydrates and uptake of macromolecules by plant cells. The role of the partially coated reticulum (PCR) and multivesicular bodies (MVB) in endocytosis is discussed by Fowke *et al.* (this volume). Although there is little evidence for a vesicle shuttle between Golgi cisternae in plants there is evidence that exit from the Golgi can be from the medial cisternae, *trans*-cisternae and from the *trans*-Golgi network (TGN) (see Staehelin *et al.*, and Kronestedt-Robards & Robards, this volume). RER, rough endoplasmic reticulum.

between the ER and the *trans*-Golgi network, a plant equivalent of GERL (Golgi–ER–lysosome system: Harris 1986; Harris & Oparka 1982).

Transport through the Golgi

The Golgi apparatus is the key organelle where much processing and sorting of biosynthetic macromolecules takes place (see Harris & Watson, this volume). At first glance it appears that the plant system may be very different from that in animal cells. At the microscopical level the Golgi apparatus appears distinct, with individual dictyosomes, and is not so interconnected and engulfed in networks of endomembranes as in animal cell systems. Perhaps this is an evolutionary result of polysaccharides being the predominant class of secreted molecules (Fig. 3). However, work carried out over the last decade has confirmed that in many respects vesicle traffic and membrane flow in plant and animal cells systems is comparable (Figs. 1 and 2). For instance it is now known that plants have a clathrin-mediated endocytotic pathway very similar to, if not identical with, that in animal cell systems (see Fowke *et al.*, Coleman *et al.* and Robinson *et al.*, this volume). The exocytotic pathway to the plasma membrane in both plants and animals is the default (not-signalled) pathway whereas the signalled (targeted) pathway is to the lysosomal compartment, although in the latter case the mode of signalling appears to have different mechanisms in plants (see Harris & Watson and Jones *et al.*, this volume) and animals (see below).

Yeast cells are becoming popular model systems in which to study secretory processes, due to the ease with which the genetic control over the pathways of vesicle traffic can be dissected. Major breakthroughs in cell biology have come from the use of temperature-sensitive mutants in yeast to reveal genes and gene products involved in cellular events which may then be used to probe their homologues in mammalian systems. Temperature-sensitive mutants are cells which function normally within a permissive temperature range, but in which the mutant genes are expressed within a restrictive temperature range to give altered non-functional gene products. Thus, mutant cells can be grown at low temperatures and simply by raising the temperature the affected gene is switched off and the mutant phenotype can be studied.

The existence of temperature-sensitive secretory (*sec*) mutants in yeast has been known for 10 years (Novick *et al.* 1981; Stevens *et al.* 1982) and it is clear that a large number of genes (at least 40, see Harris and Watson, this volume) are involved in the secretory process from synthesis in the ER to final vesicle fusion with the plasma membrane or vacuolar

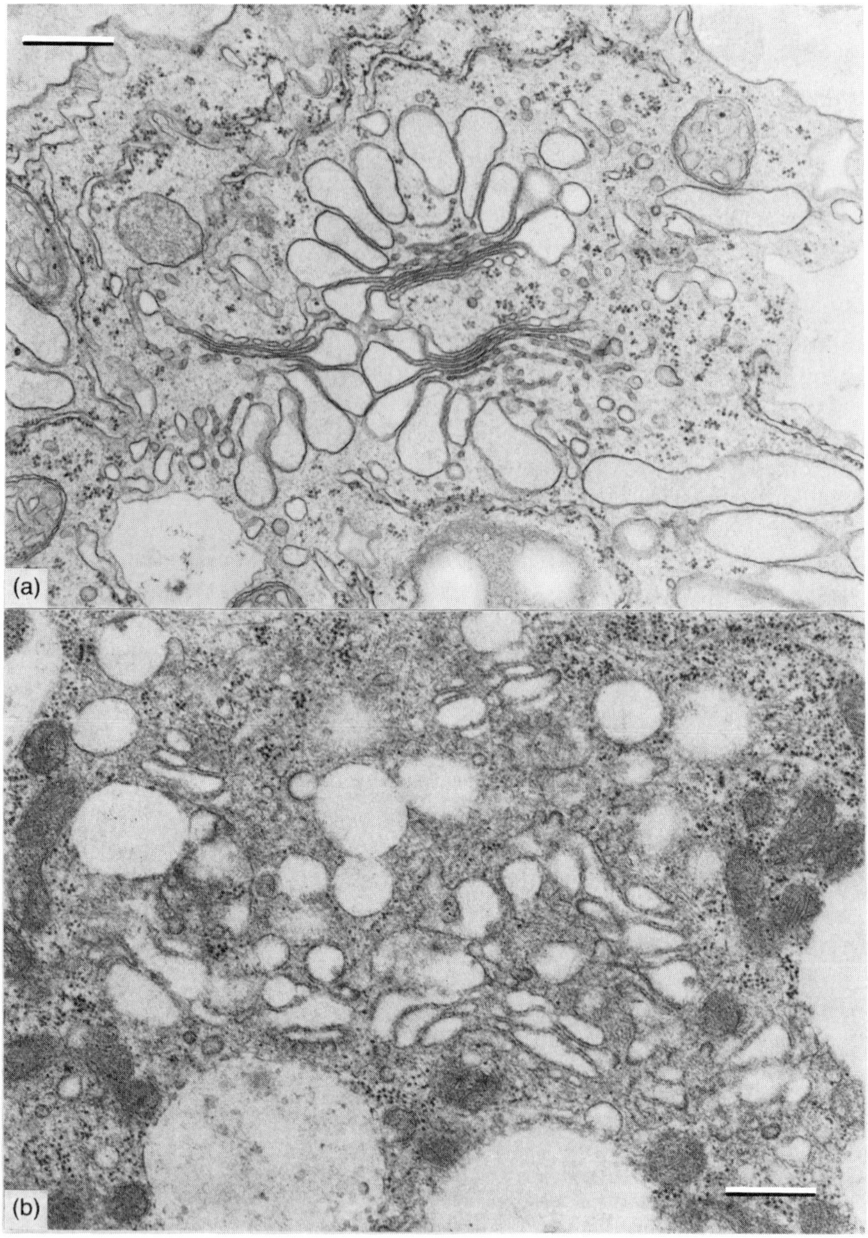

Fig. 3. Comparison between a plant carbohydrate-secreting Golgi body (a) maize root cap cell and an animal cell Golgi apparatus (b) 3T3 fibroblast. The bars represent: (a) 0.8 μm, (b) 0.5 μm.

membrane. A recent series of papers has revealed that gene products which operate in the early stages of the yeast secretory pathway (Kaiser & Schekman 1990) are the functional homologues of recently identified proteins involved in Golgi vesicle shuttles and in vesicle/membrane docking and fusion in general (Wilson *et al.* 1989; Clary *et al.* 1990).

Studies on the dynamics and molecular mechanisms of the shuttle of Golgi cisternal vesicles down the stack, carried out predominantly in Rothman's laboratory, have shed new light on possible mechanisms of vesicle–membrane docking and fusion as well as on the operation of the Golgi itself. Initial work assaying the transport of a glycoprotein of the vesicular stomatitis virus in a cell-free system from Chinese hamster ovary (CHO) cells showed that the transfer of vesicles between the ER and Golgi and down the Golgi stack required ATP, fatty acyl CoA and at least two cytosolic proteins (Orci *et al.* 1989). These vesicles were shown to be coated, but with a protein that was not clathrin (Orci *et al.* 1986). Treatment of the system with the GTP analogue GTP-γ-S resulted in an accumulation of these coated vesicles around the Golgi stack. Thus, it was concluded that one of the cytosolic proteins was a GTP-binding protein required for uncoating the vesicles prior to their docking and fusion with the next cisternum.

More recently a series of proteins has been identified and implicated in vesicle docking and fusion process. The first, NSF (*N*-ethylmaleimide-sensitive factor), is involved in the fusion process after the uncoated vesicle has been bound to the docking membrane (Block *et al.* 1988). NSF will apparently bind only to isolated membranes in the presence of a series of three proteins collectively termed SNAP (soluble NSF attachment proteins) (Weidman *et al.* 1989). It is thought that these proteins link the NSF directly to a receptor on the Golgi cisternum. Another protein of 25 000 M_r, in this sequence of events, which is required at the last stage of vesicle/membrane fusion, has just been isolated (Wattenberg *et al.* 1990). Termed POP (protein acting in prefusion), it has been assigned several putative functions such as being a genuine fusogen, directing proteins to the sites of fusion, or unmasking fusion sites on the acceptor membrane. However, what is exciting is that this protein was isolated from yeast cytosol and mediated vesicle fusion in the CHO cell-free assay system. Of equal importance are the recent reports that the yeast *sec18* gene product is functionally and structurally homologous to NSF (Wilson *et al.* 1989) and the *sec17* gene product is α-SNAP (Clary *et al.* 1990). Fusion of transition vesicles between the ER and Golgi (Beckers *et al.* 1989) and of endocytotic vesicles has also been shown to be dependent on NSF (Diaz *et al.* 1989).

Thus, it is rapidly becoming apparent that there may be one universal system mediating fusion of vesicles to acceptor membranes. Surely, it will not be long before functional homologues of these proteins are characterised in higher plant systems.

Exit from the Golgi and post Golgi pathways

As mentioned above it is becoming apparent that in all eukaryotic cells vesicular transport to the plasma membrane is along a default pathway whilst transport to the lysosomal/vacuolar compartment is the targeted/signalled pathway. The addition of targeting information to biosynthetic macromolecules is carried out in the Golgi apparatus and the products sorted in the *trans*-Golgi network. Perhaps the best characterised signalled system is that in animal cells, which is based on the addition of mannose-6-phosphate to lysosomal enzymes that are transported to the early lysosome (Brown *et al.* 1986). Within the *cis*-Golgi a mannose-6-phosphate group is added to the *N*-linked oligosaccharides of glycoproteins and in the *trans*-Golgi network these groups bind to receptors and cluster into clathrin-coated pits. The receptors apparently link to the clathrin triskelions via specific adaptor proteins (Morris *et al.* 1989). The vesicles pinch off from the Golgi, uncoat and are transported to the early lysosome/late endosome where, due to an acidification of the lysosomal lumen, receptor and ligand part company and the receptor is recycled, via another vesicle to the *trans*-Golgi network. Thus, a selective mechanism is in operation which is similar to that of receptor-mediated endocytosis at the plasma membrane.

Important questions then have to be asked. What is the function of the plant Golgi associated clathrin-coated vesicles? To what extent are they involved in targeted protein transport and if so what are the receptors and targeting mechanisms? Some preliminary evidence exists which suggests that they may be involved in protein transport from the Golgi (see Robinson *et al.*, this volume; Harley & Beevers 1989), but much more work remains to be carried out. Techniques will have to be developed to separate reliably Golgi from plasma-membrane-derived coated vesicles before fully satisfactory biochemical analyses can be carried out.

To date, the only information on signalling in the targeted pathway for both yeasts and plants lies in the sequence of the N-terminal amino acid residues of the transported protein. For instance, in yeast cells, if the secreted enzyme invertase has tagged on to it by genetic manipulation the 30 N-terminal amico acid residues of the precursor of the vacuolar enzyme carboxypeptidase Y, the hybrid protein is targeted to the vacuole (see Rothman *et al.* 1989, for a review on protein targeting to the yeast vacuole). In plants it is known that tobacco transformed with phaseolin and phytohaemagglutinin genes from bean can target these bean seed storage proteins to vacuolar protein bodies (Greenwood & Chrispeels 1985; Herman *et al.* 1989). Apparently, glycosylation signals do not operate in bean seed protein targeting (Voelker *et al.* 1989). However, the same group have come to the conclusion that more than a signal peptide alone is required for correct targeting of proteins to a storage vacuole (Dorel *et al.* 1989). In contrast the default pathway appears to be in operation as far back in the system as the lumen of the ER. Proteins not normally secreted are actively secreted if targeted to the lumen of the ER by signal-peptide-mediated translocation (Denecke *et al.* 1990).

Although it is generally accepted that in all cells the default pathway of secretion is to the plasma membrane, in animal cell systems this is by no means uncontrolled. There are two distinct forms of exocytosis, constitutive and regulated (see Huttner & Tooze 1989, for a review). In the former, biosynthesised molecules are continually produced and secreted by the cell as Golgi-derived vesicles which fuse with the plasma membrane. However, this system is not in all cases devoid of signals. The majority of cells in animal tissues are polarised and may have distinct apical and basolateral domains such as are exhibited by epithelia. Such cells can secrete different sets of proteins to different surfaces, thus sorting and/or directional signals must exist. Clues to such signals have been derived from the study of enveloped viruses (Roman & Garoff 1985). The viral envelope, with its characteristic set of proteins, is budded off from the Golgi and inserted into the plasma membrane along with the nucleocapsid. In a polarised sheet of cells, the influenza virus will bud only from the apical domain whilst the Semliki forest and vesicular stomatitis viruses bud only from the basolateral plasma membranes (Rodriguez-Boulan & Sabatini 1978), the signalling information being in the proteins of the virus envelope.

Regulated secretion occurs in specialised secretory cells that need to release their products rapidly on demand, normally in response to an extracellular signal. This is often accompanied by a transient rise in cytosolic [Ca^{2+}], which promotes fusion of the vesicle membrane with the plasma membrane. A classic example of this system is the cortical reaction during echinoderm fertilisation, where the interaction of the acrosomal vesicle with the vitelline layer initiates the inositol phospholipid signalling pathway, resulting in a release of the free Ca^{2+} into the cytosol and a massive fusion of cortical granules with the egg plasma membrane (Whitaker & Irvine 1984).

There is no hard evidence that such a mechanism for regulated secretion exists in higher plant cells, although examples can be found where exocytosis is stimulated by an extracellular signal. Classic cases of this would be the exocytosis of digestive enzymes after stimulation of glands on the leaves of insectivorous plants (Outenreath & Dauwalder 1986) or the release of α-amylase by cereal aleurone cells after stimulation with gibberellic acid (see Jones *et al.*, this volume).

Endocytotic pathways

The pathways of endocytosis and especially the process of receptor-mediated endocytosis and the subsequent processing of receptors and ligands are perhaps the best characterised of all the vesicle trafficking systems. However, the available data have been obtained almost exclusively from animal cell systems (van Deurs *et al.* 1989). Experiments on yeast cells have been restricted to fluorescent studies on endocytosis (Riezman 1985; Riezman *et al.* 1986), the biochemical extraction of clathrin (Mueller & Branton 1984) and some work on clathrin-free mutants (Payne & Schekman 1985). Problems with satisfactory ultrastructural preservation of yeast cells have to date precluded the study of the endocytotic pathway with electron-opaque probes.

The role of clathrin in the formation of coated pits and vesicles at the plasma membrane is well characterised (for a review, see Brodsky 1988) as is the fate of the endocytosed vesicle after uncoating and fusion with the early endosome (see Fig. 1; and for a review, Hubbard 1989). New information is beginning to emerge on the mechanics of clathrin binding via adaptor proteins to the cytoplasmic tails of plasma membrane-bound receptors (Morris *et al.* 1989; see also Coleman *et al.*, this volume). The full mode of action of a universal binding system to cross-link clathrin to a heterogeneous group of receptor peptides has yet to be fully elucidated.

Endocytosis is involved not only in the uptake of important macromolecules, but is an event that may be of great importance in maintaining the membrane economy of the cell. In a non-growing cell, exocytosis will have to be balanced by endocytosis to maintain the *status quo* of the plasma membrane (see Steer & O'Driscoll, this volume). Also, in motile animal cells endocytosis has been implicated in the recycling of the plasma membrane as it flows backwards over the surface of the cell away from the extending lamellipodia of the growing end (Bretscher 1987).

It is now clear (Coleman *et al.* 1988, and this volume) that the elements of receptor-mediated endocytosis exist in plants. However, specific plasma membrane receptors for internalised molecules have yet to be characterised, and the identification of the endocytotic pathway has come only from studies of the internalisation of tracer molecules.

From this brief review it can be seen that research carried out during the last decade has resulted in major advances in our understanding not only of the biosynthetic machinery of cells but also of the mechanisms by which cells regulate the transport of such products to the right place at the right time. Concurrently the cell can internalise molecules and also incorporate them into the complex system of endomembranes and compartments.

Acknowledgement

We thank Mr Derek Whiteley of Oxford Polytechnic for his help in drawing Figs. 1 and 2.

References

Beckers, C.J.M., Block, M.R., Glick, B.S., Rothman, J.E. & Balch, W.E. (1989). Vesicular transport between the endoplasmic reticulum and the Golgi stack requires the NEM-sensitive fusion protein. *Nature* (*London*), **339**, 397–400.

Block, M.R., Glick, B.S., Wilcox, C.A., Wieland, F.T. & Rothman, J.E. (1988). Purification of an *N*-ethylmaleimide-sensitive (NSF) protein catalyzing vesicular transport. *Proceedings of the National Academy of Sciences, USA*, **85**, 7852–6.

Bretscher, M.S. (1987). How animal cells move. *Scientific American*, **257**, 44–50.

Brodsky, F.M. (1988). Living with clathrin: its role in intracellular membrane traffic. *Science*, **242**, 1396–402.

Brown, W.J., Goodhouse, J. & Farquhar, M.G. (1986). Mannose-6-phosphate receptors for lysosomal enzymes cycle between the Golgi complex and endosomes. *Journal of Cell Biology*, **103**, 1235–47.

Clary, D.O., Griff, I.C. & Rothman, J.E. (1990). SNAPs, a family of NSF attachment proteins involved in intracellular membrane fusion in animals and yeast. *Cell*, **61**, 709–21.

Coleman, J., Evans, D. & Hawes, C.R. (1988). Plant coated vesicles. *Plant, Cell and Environment*, **11**, 669–84.

Dean, N. & Pelham, H.R.B. (1990). Recycling of proteins from the Golgi compartment to the ER in yeast. *Journal of Cell Biology*, **111**, 369–77.

Denecke, J., Botterman, J. & Deblaere, R. (1990). Protein secretion in plants can occur via a default pathway. *Plant Cell*, **2**, 51–9.

Diaz, R., Mayorga, L.S., Weidman, P.J., Rothman, J.E. & Stahl, P.D. (1989). Vesicle fusion following receptor mediated endocytosis requires a protein active in Golgi transport. *Nature* (*London*), **339**, 398–400.

Dorel, C., Voelker, T.A., Herman, E.M. & Chrispeels, M.J. (1989). Transport of proteins to the plant vacuole is not by bulk flow through the secretory system, and requires positive sorting information. *Journal of Cell Biology*, **108**, 327–37.

Greenwood, J.S. & Chrispeels, M.J. (1985). Correct targeting of the bean storage protein phaseolin in the seeds of transformed tobacco. *Plant Physiology*, **79**, 65–71.

Harley, S.M. & Beevers, L. (1989). Coated vesicles are involved in the transport of storage proteins during seed development in *Pisum sativum* L. *Plant Physiology*, **91**, 674–8.

Harris, N. (1986). Organization of the endomembrane system. *Annual Review of Plant Physiology*, **37**, 73–92.

Harris, N. & Oparka, K.J. (1982). Connections between dictyosomes, ER and GERL in cotyledons of mung bean (*Vigna radiata* L.). *Protoplasma*, **114**, 93–102.

Herman, E.M., Chrispeels, M.J. & Hoffman, L.M. (1989). Vacuole accumulation of storage protein and lectin expressed in transgenic tobacco seeds. *Cell Biology International Reports*, **13**, 37–45.

Hubbard, A.L. (1989). Endocytosis. *Current Opinion in Cell Biology*, **1**, 675–83.

Huttner, W.B. & Tooze, S.A. (1989). Biosynthetic protein transport in the secretory pathway. *Current Opinion in Cell Biology*, **1**, 648–54.

Juniper, B.E., Hawes, C.R. & Horne, J.C. (1982). The relationships between the dictyosomes and the forms of endoplasmic reticulum in plant cells with different export programmes. *Botanical Gazette*, **143**, 135–45.

Kaiser, C.A. & Schekman, R. (1990). Distinct sets of *SEC* genes govern transport vesicle formation and fusion early in the secretory pathway. *Cell*, **61**, 723–33.

Lippincott-Schwartz, J., Donaldson, J.G., Schweizer, A., Berger, E.G., Hauri, H-P., Yuan, L.C. & Klausner, R.D. (1990). Microtubule-dependent retrograde transport of proteins into the ER in the presence of Brefeldin A suggests an ER recycling pathway. *Cell*, **60**, 821–36.

Morris, S.A., Ahle, S. & Ungewickell, E. (1989). Clathrin coated vesicles. *Current Opinion in Cell Biology*, **1**, 684–90.

Mueller, S. & Branton, D. (1984). Identification of coated vesicles in *Saccharomyces cerevisiae*. *Journal of Cell Biology*, **98**, 341–6.

Munro, S. & Pelham, H.R.B. (1987). A C-terminal signal prevents secretion of luminal ER proteins. *Cell*, **48**, 899–907.

Novick, P., Ferro, S. & Schekman, R. (1981). Order of events in the yeast secretory pathway. *Cell*, **25**, 461–9.

Orci, L., Glick, B.S. & Rothman, J.E. (1986). A new type of coated vesicle carrier that appears not to contain clathrin: its possible role in protein transport within the Golgi stack. *Cell*, **46**, 171–84.

Orci, L., Malhotra, V., Amherdt, M., Serafini, T. & Rothman, J.E. (1989). Dissection of a singular round of vesicular transport: sequen-

tial intermediates for intercisternal movement in the Golgi stack. *Cell*, **56**, 357–68.
Outenreath, R.L. & Dauwalder, M. (1986). Ultrastructural and radioautographic studies of the digestive gland cells of *Drosera capensis*. *Journal of Ultrastructure and Molecular Structure Research*, **95**, 164–74.
Payne, G.S. & Schekman, R. (1985). A test of clathrin function in protein secretion and cell growth. *Science*, **230**, 1009–14.
Pelham, H.R.B. (1988). Evidence that luminal ER proteins are sorted from secreted proteins in a post-ER compartment. *EMBO Journal*, **7**, 913–18.
Pelham, H.R.B., Hardwick, K.J. & Lewis, M.J. (1988). Sorting of soluble ER proteins in yeast. *EMBO Journal*, **7**, 1757–62.
Riezman, H. (1985). Endocytosis in yeast: several of the yeast secretory mutants are defective in endocytosis. *Cell*, **40**, 1001–9.
Riezman, H., Chvatchko, Y. & Dulic, V. (1986). Endocytosis in yeast. *Trends in Biochemical Sciences*, **11**, 325–8.
Rodriguez-Boulan, E.J. & Sabatini, D.D. (1978). Asymmetric budding of viruses in epithelial monolayers: a model system for the study of epithelial polarity. *Proceedings of the National Academy of Sciences, USA*, **75**, 5071–5.
Roman, L.M. & Garoff, H. (1985). Revelation through exploitation: the viral model for intracellular traffic. *Trends in Biochemical Sciences*, **10**, 428–32.
Rothman, J.H., Yamashiro, C.T, Kane, P.M. & Stevens, T.H. (1989). Protein targeting to the yeast vacuole. *Trends in Biochemical Sciences*, **14**, 347–50.
Semenza, J.C., Hardwick, K.G., Dean, N. & Pelham, H.R.B. (1990). *ERD2*, a yeast gene required for the receptor-mediated retrieval of luminal ER proteins from the secretory pathway. *Cell*, **61**, 1349–57.
Stevens, T., Esmon, B. & Schekman, R. (1982). Early stages in the yeast secretory pathway are required for transport of carboxypeptidase Y to the vacuole. *Cell*, **30**, 439–48.
van Deurs, B., Petersen, O.W., Olsnes, S. & Sandvig, K. (1989). The ways of endocytosis. *International Review of Cytology*, **117**, 131–77.
Vaux, D., Tooze, J. & Fuller, S. (1990). Identification by anti-idiotype antibodies of a intracellular membrane protein that recognizes a mammalian endoplasmic retention signal. *Nature (London)*, **345**, 495–502.
Voelker, T., Herman, E. & Chrispeels, M.J. (1989). *In vitro* mutated phytohemagglutinin genes expressed in tobacco seeds: role of glycans in protein targeting and stability. *Plant Cell*, **1**, 95–104.
Wattenberg, B.W., Hiebsch, R.R., LeCureux, L.W. & White, M.P. (1990). Identification of a 25-kD protein from yeast cytosol that operates in a prefusion step of vesicular transport between compartments of the Golgi. *Journal of Cell Biology*, **110**, 947–54.

Weidman, P. J., Melançon, P., Block, M. R. & Rothman, J. E. (1989). Binding of an *N*-ethylmaleimide-sensitive fusion protein to Golgi membranes requires both a soluble protein(s) and an integral membrane receptor. *Journal of Cell Biology*, **108**, 1589–96.

Whitaker, M. J. & Irvine, R. F. (1984). Inositol, 1,4,5-trisphosphate microinjection activates sea urchin eggs. *Nature* (*London*), **308**, 636–9.

Wilson, D. W., Wilcox, C. A., Flynn, G. C., Chen, E., Kuang, W-J., Henzel, W. J., Block, M. R., Ullrich, A. & Rothman, J. E. (1989). A fusion protein required for vesicle mediated transport in both mammalian cells and yeast. *Nature* (*London*), **339**, 355–9.

Notes added in proof

[1] The amino acid sequence KDEL also acts as an ER retention signal in higher plants (Inohara *et al.* (1989), *Proceedings of the National Academy of Sciences, USA* **86**, 3564–8; Herman *et al.* (1990), *Planta* **182**, 305–312

[2] Since this chapter was written a series of papers has been published on the structure of Golgi-derived coated vesicles. Four high molecular weight proteins form the main subunits of the non-clathrin Golgi-derived coated vesicles (COPs) in CHO cells (Serafini *et al.* (1991), *Nature* (*London*) **349**, 215–19). These proteins reside in a cytosolic complex termed a 'coatomer' (Waters *et al.* (1991), *Nature* (*London*), **349**, 248–51). One of these proteins β-COP, 110×10^3 M_r, has been shown to have close sequence homology with β-adaptin, which is present in both Golgi and plasma membrane clathrin adaptor comlexes (Duden *et al.* (1991), *Cell*, **64**, 649–65).

L.C. FOWKE, M.A. TANCHAK AND M.E. GALWAY

Ultrastructural cytology of the endocytotic pathway in plants

In recent years studies of endocytosis in animal cells have focused largely on the highly specific mechanism of receptor-mediated endocytosis which internalises a range of protein molecules. These include transport proteins such as the cholesterol-carrying protein low-density lipoprotein (LDL) and the iron transport protein transferrin, the non-transport proteins asialoglycoprotein and immune complexes, lysosomal enzymes, protein growth factors and polypeptide hormones such as insulin (e.g. Pearse & Bretscher 1981; Fine & Ockleford 1984; Goldstein *et al.* 1985; Wileman *et al.* 1985). These proteins are bound to receptors in the plasma membrane, endocytosed by clathrin-coated pits to coated vesicles and then delivered to endosomes, acidic compartments where the proteins are dissociated from the receptors. Some proteins (e.g. transferrin) are returned to the cell surface while others (e.g. immunoglobulin A) are transported across the cell for secretion. Many proteins (e.g. LDL, insulin), however, are delivered to the lysosome for degradation by hydrolytic enzymes. Ultrastructural studies have played a key role in defining the process of receptor-mediated endocytosis in animal cells.

Until recently, little was known about either receptor-mediated endocytosis or endocytosis in general in plant cells. Initially endocytosis by plant cells was believed to be unlikely because of turgor pressure (Cram 1980), but recent reports indicate that endocytosis via coated pits and coated vesicles is energetically possible (Saxton & Breidenbach 1988), particularly in cells with relatively low turgor pressure, such as root hairs (Gradman & Robinson 1989). Indeed, in the literature, there are many reports of uptake into cells of plant roots (for a discussion, see Coleman *et al.* 1988). The results of Hübner *et al.* (1985) suggest that coated membranes are involved in this type of uptake in root cap cells of maize. Recent experiments utilising membrane-impermeant fluorescent probes clearly demonstrate that plant cells are capable of endocytosis (see e.g.

Oparka *et al.* 1988; Hillmer *et al.* 1989). Analysis of the energetics of internalisation by barley roots suggests that fluid phase endocytosis provides a mechanism for this uptake (Oparka *et al.* 1988).

Plant protoplasts (wall-less plant cells) have been used extensively for studies of endocytosis. Although atypical with respect to morphology and physiology, they provide a unique experimental system for examining events at the plasma membrane surface during endocytosis. Uptake experiments can be designed using a variety of probes, many of which would not penetrate the intact cell wall. Early studies with protoplasts detailed uptake of various materials, including intact cells and even other protoplasts (for reviews, see Fowke & Gamborg 1980; Fowke 1986) but shed very little light on mechanisms of internalisation. For example, polystyrene spheres and bacteria with or without cell walls were taken into protoplasts by invagination of uncoated regions of the plasma membrane. Uptake of the polystyrene spheres was an energy-dependent process. Larger membrane-bounded structures such as unicellular algal protoplasts were taken up by fusion with the larger host protoplast rather than by endocytosis.

Research with protoplasts during the past six years has provided new and exciting information concerning uptake of macromolecules by coated membrane-mediated endocytosis and fluid phase endocytosis in plants (Tanchak *et al.* 1984; Joachim & Robinson 1984; Hillmer *et al.* 1986; Tanchak & Fowke 1987; Griffing 1988; Wright & Oparka 1989). Ultrastructural studies of the uptake of electron-dense markers have been confined to protoplasts from soybean cells (see below) and from bean leaf and carrot cells (Joachim & Robinson 1984; Hillmer *et al.* 1986).

Organelles involved in endocytosis

Electron microscope studies have identified a number of organelles involved in endocytosis in plants. A brief description of these organelles will facilitate subsequent interpretation of experimental results illustrating an endocytotic pathway in plant cells.

Coated pits and coated vesicles

Coated pits and coated vesicles have been described in a broad spectrum of plants including algae, simple land plants and vascular plants (Newcomb 1980). A variety of electron microscope techniques have been used to investigate their morphology. These include serial thin sectioning (Fowke *et al.* 1989), negative staining (van der Valk & Fowke 1981), and freeze etching (Hawes & Martin 1986; Coleman *et al.* 1988, and this volume). Clathrin-coated regions of the plasma membrane invaginate as coated pits which are released into the cytoplasm as coated vesicles. In thin sections, the clathrin coat is evident as bristles extending into the cytoplasm (Figs. 1 to 3). The characteristic surface pattern of the clathrin coat consisting of hexagons and pentagons is clearly recognisable in sections (Fig. 3), stained plasma membrane preparations (Fig. 4) and on isolated coated vesicles (Fig. 5). Coated pits and coated vesicles in plants closely resemble their counterparts in animal cells (see e.g. Ockleford & Whyte 1980).

Coated pits and coated vesicles are particularly numerous in cells actively depositing new cell wall material (e.g. root hairs, Robertson & Lyttleton 1982; cotton fibres, Ryser 1979) and at the cell plate region of dividing plant cells (see Fig. 6; and Franke & Herth 1974; Nakamura & Miki-Hirosige 1982). Using techniques which permit direct examination of the inner surface of the plasma membrane of protoplasts (van der Valk *et al.* 1980) and intact plant cells (Traas 1984), it has been possible to illustrate a positive correlation between active growth involving cell wall deposition and the frequency of coated pits. The plasma membrane of tobacco protoplasts initiating cell wall regeneration showed an eight-fold increase in coated pits compared to membranes of protoplasts unable to form walls (Figs. 7 and 8, and Fowke *et al.* 1983*b*). Emons & Traas (1986) examined the plasma membranes of various plant cells and observed a similar correlation.

Partially coated reticulum (PCR)

The PCR is a system of interconnected tubular membranes bearing coated regions and often surrounded by coated vesicles (Figs. 9 and 10). It was first described by Pesacreta & Lucas (1984) in the alga *Chara* and has since been reported in cells and protoplasts of various angiosperms (see e.g. Pesacreta & Lucas 1985; Tanchak *et al.* 1984; Hillmer *et al.* 1986). Following high-pressure freezing and freeze substitution (Hippe *et al.* 1989), the morphology of the PCR appeared to be the same as that achieved with conventional chemical fixation. In soybean protoplasts the PCR occasionally appeared in a ring configuration and was often characterised by a rounded membrane dilation containing small internal vesicles (Fig. 10). The idea that the PCR is distinct from the Golgi apparatus (Golgi) is based primarily on serial sectioning analyses which have shown that, while the PCR is often closely associated with Golgi, very rarely is there a direct connection between the two structures (Pesacreta & Lucas 1985; Tanchak *et al.* 1988). In the study of Tanchak *et al.* (1988), the PCR was clearly labelled with endocytosed cationised ferritin (see below) thus facilitating identification of this structure from section to section. Studies using material stained with zinc iodide–osmium tetroxide further

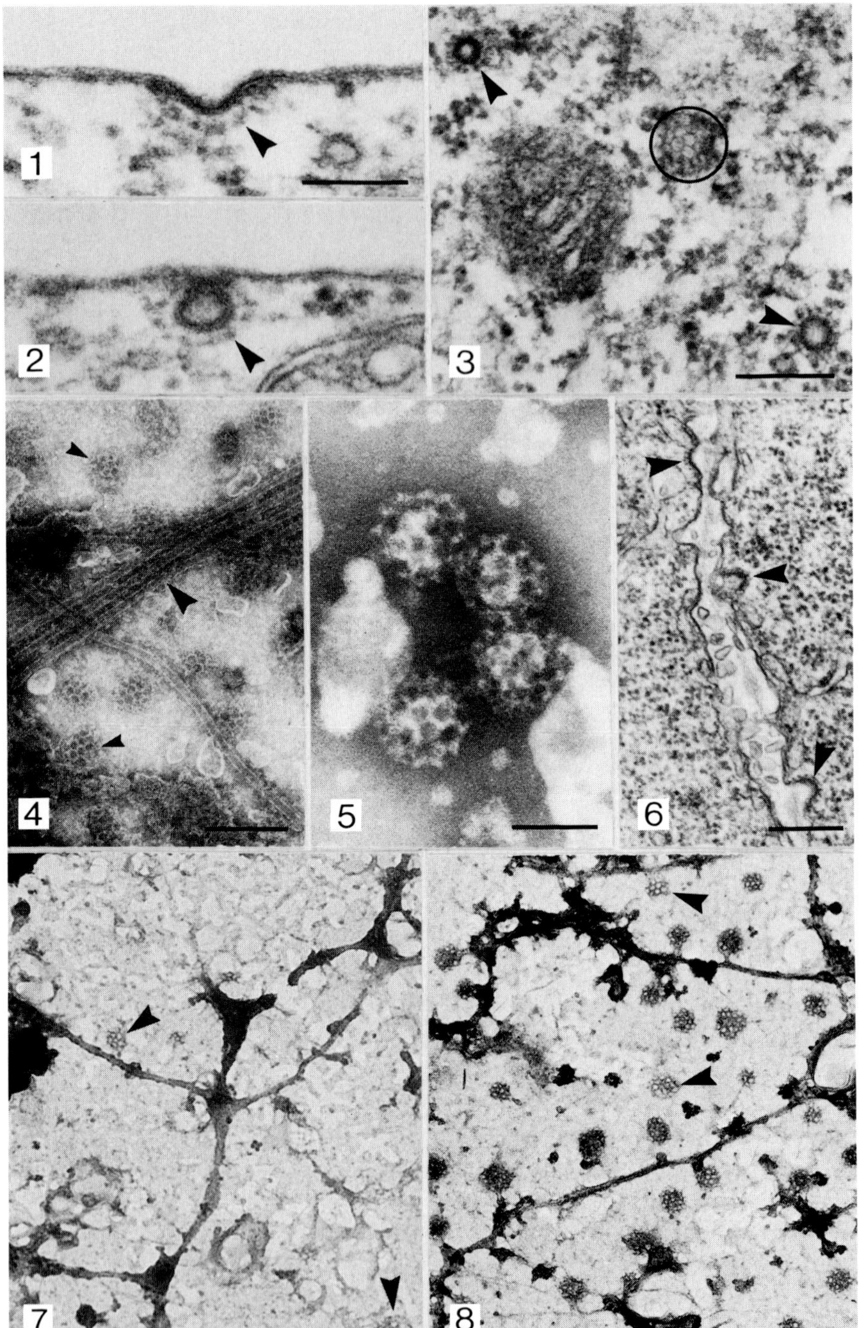
1
2
3
4
5
6
7
8

differentiated between the PCR and Golgi. In sections, the PCR was lightly stained and could be distinguished clearly from adjacent heavily stained Golgi (Pesacreta & Lucas 1985; Tanchak 1987). Furthermore, Record & Griffing (1988) demonstrated the presence of acid phosphatase in Golgi of soybean protoplasts but not in the PCR. Hillmer *et al.* (1988) in describing the close relationship between the PCR and Golgi in various plant cells hypothesised that the PCR is derived physically from Golgi by transcisternal release. However, they offered little supporting evidence for this idea.

Golgi

Plant Golgi consist of stacks of flattened cisternae with fenestrated edges (see Robinson & Kristen 1982). They occur singly in the cytoplasm and unlike animal Golgi they seldom aggregate to form a complex apparatus.

Figs. 1–8. (1) Early stage of coated pit formation in tobacco protoplast. The clathrin coat (arrowhead) is clearly visible on the cytoplasmic surface of the plasma membrane. The bar represents 100 nm. (From van der Valk & Fowke 1981.)

(2) Late stage of coated pit formation in tobacco protoplast. Note the clathrin coat (arrowhead). Same magnification as Fig. 1. (From van der Valk & Fowke 1981.)

(3) Cytoplasm of tobacco protoplast showing sectioned coated vesicles (arrowheads) and surface structure of a coated pit or coated vesicle (encircled). The bar represents 200 nm. (From van der Valk & Fowke 1981.)

(4) Negatively stained inner surface of tobacco protoplast plasma membrane showing surface features of coated pits (small arrowheads) and microtubules (large arrowhead). The bar represents 200 nm.

(5) Negatively stained coated vesicles isolated from suspension cultured tobacco cells. The bar represents 100 nm.

(6) Region of cell plate from dividing *Marchantia* cell showing coated pits (arrowheads). The bar represents 200 nm. (From Fowke & Pickett-Heaps 1978.)

(7) Inner surface of plasma membrane from tobacco leaf protoplast, which does not deposit new cell wall until after a lag period. Such protoplasts have a low frequency of coated pits (arrowheads). The bar represents 300 nm.

(8) Inner surface of plasma membrane of protoplast from rapidly growing cultured tobacco cells capable of immediate cell wall deposition. Such protoplasts have numerous coated pits (arrowheads). Same magnification as Fig. 7.

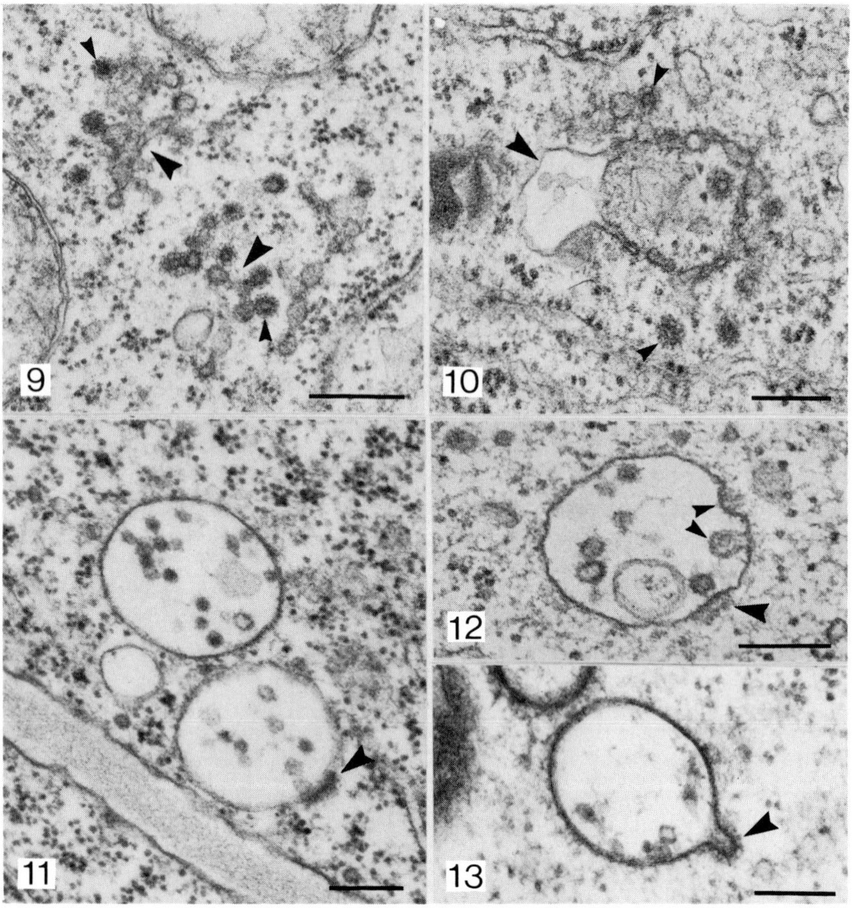

Figs. 9–13. (9) Two PCRs (large arrowheads) in a somatic embryo of white spruce. Note the numerous coated regions (small arrowheads). The bar represents 300 nm. (From Hakman *et al.* 1987.)
(10) PCR in soybean protoplast. Note the ring configuration and membrane dilation (large arrowhead) containing vesicles. Coated vesicles (small arrowheads) are present near the PCR. The bar represents 200 nm. (From Tanchak *et al.* 1988.)
(11) Two MVBs in a somatic embryo of white spruce. A plaque (arrowhead) is present on one MVB. The bar represents 200 nm.
(12) MVB in soybean protoplast. One surface plaque shows structural periodicity (large arrowhead). Internal vesicles appear to arise by an invagination process (small arrowheads). The bar represents 200 nm.
(13) MVB in soybean protoplast. Note the coated pit (arrowhead) on the surface of the MVB. The bar represents 200 nm. (From Tanchak & Fowke 1987.)

Multivesicular bodies (MVBs)

MVBs were described in the 1960s in a variety of plants and were initially believed to be involved in the process of cell wall deposition (for a review, see Marchant & Robards 1968), although direct evidence was not available to support this contention. Interest in MVBs in plants dwindled until 1984 when they were first implicated in endocytosis (Tanchak *et al.* 1984). MVBs are single membrane-bounded organelles, approximately 250–500 nm in diameter, containing numerous small membraneous vesicles (Figs. 11 to 13). Distinctive plaques, some with a structural periodicity similar to clathrin coats, are frequently present on the outer surface (Fig. 12). Coated pits are occasionally observed on MVBs (Fig. 13) or at the ends of protuberances extending from the MVB surface (Hillmer *et al.* 1986). MVBs are distributed throughout the cytoplasm but are preferentially located in the perinuclear region of cells. They may contain acid phosphatase and/or peroxidase (Halperin 1969; Buvat & Robert 1979; Griffing & Fowke 1985; Record & Griffing 1988).

Plant MVBs are similar to MVBs in animal cells in terms of size, morphology, distribution, enzyme complement and reaction to zinc iodide–osmium tetroxide (for a more complete discussion see Tanchak & Fowke 1987). MVBs in animal cells are believed to represent late endosomes in terms of processing endocytosed ligands, and are thought to fuse ultimately with lysosomes. The fact that they contain very few hydrolytic enzymes indicates that they may not themselves be true lysosomes. Kornfeld & Mellman (1989) argue that the term lysosome should be reserved only for the terminal degradative compartment in cells containing the full complement of lysosomal enzymes.

Vacuoles

Mature plant cells are characterised by a large central vacuole and numerous small vacuoles in the peripheral cytoplasm. Vacuoles have a variety of functions in plants including storage (amino acids, sugars), waste disposal (crystals, organic acids) and turgor-related phenomena such as cell expansion and the movement of ‘sensitive plants’ (e.g. *Mimosa*) (Gunning & Steer 1975). In addition, there is considerable evidence, particularly from studies of plant protoplasts, that plant vacuoles contain a range of hydrolytic enzymes and thus represent the lytic compartment or lysosome of plants (Matile 1978; Boller & Kende 1979; Nishimura & Beevers 1979). The possibility exists that vacuoles within individual cells are specialised to accommodate this range of functions. Indeed there is evidence indicating that plant cells contain at least two different types of vacuole (e.g. Guilliermond 1941; Griffing & Fowke 1985).

Endocytosis in soybean protoplasts

Protoplasts of soybean (Fig. 14) have been thoroughly studied using both light and electron microscopy and thus serve as a model system for examining mechanisms of endocytosis in plants. The following discussion focuses on ultrastructural results which demonstrate that a non-specific marker, cationised ferritin (CF), is internalised by coated pits to coated vesicles and subsequently delivered to cytoplasmic organelles in soybean protoplasts. The results were obtained from experiments in which protoplasts were exposed to CF and fixed for electron microscopy at various times between 0 and 3 h after labelling. By examining changes in the intracellular distribution of CF at various times after labelling, a time course of the uptake and movement of CF within the protoplasts can be constructed. In order to recognise CF particles clearly during uptake and transport within protoplasts, only unstained or very lightly stained sections were used. Thus many of the micrographs demonstrating endocytosis are of low contrast (e.g. Figs. 15 to 21, 23 (a) and (b)).

Figs. 14–21. (14) Freshly isolated soybean protoplasts. The bar represents 20 μm.
(15) Internalisation of CF by coated pit (arrowhead) in soybean protoplast exposed to CF for 30 s. The bar represents 100 nm.
(16) CF in coated vesicle (arrowhead) in soybean protoplast exposed to CF for 30 s. Same magnification as Fig. 15.
(17) PCR in soybean protoplast exposed to CF for 2 min. Note the CF in coated regions (arrowheads). The bar represents 200 nm. (From Tanchak *et al.* 1988.)
(18) PCR in soybean protoplast exposed to CF for 60 min. The tubules (small arrowheads) are heavily labelled but the membrane dilation (large arrowhead) is not. The bar represents 200 nm.
(19) Golgi stack in soybean protoplast exposed to CF for 60 min. CF is present at the ends of cisternae (arrowheads). The bar represents 200 nm.
(20) MVB in soybean protoplast exposed to CF for 12 min. A small number of CF particles (arrowhead) are attached to internal vesicles. The bar represents 100 nm.
(21) MVB in soybean protoplast exposed to CF for 60 min. Numerous CF particles are attached to internal vesicles (arrowheads) and the inner surface of the MVB membrane. Same magnification as Fig. 20. (From Tanchak & Fowke 1987.)

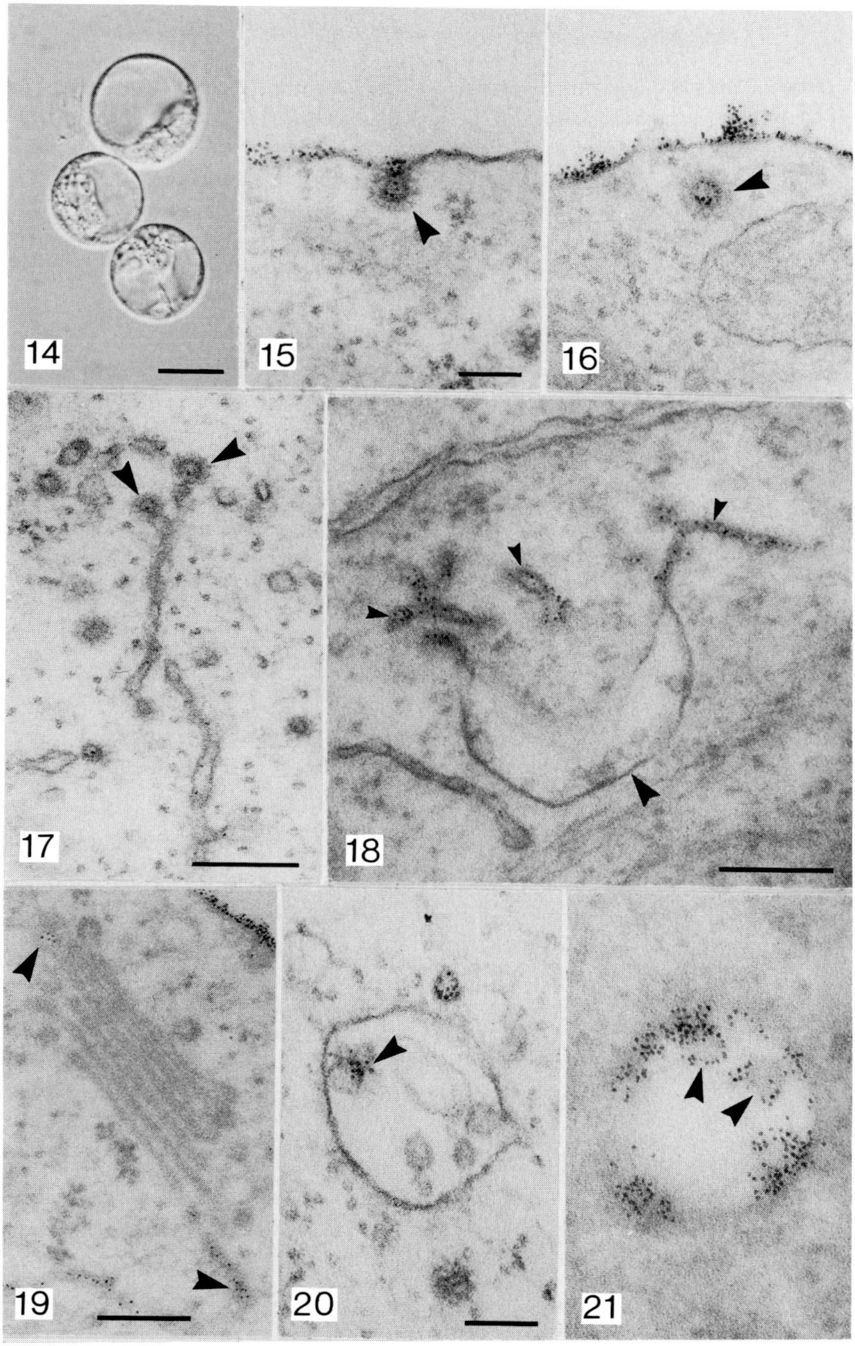
14
15
16
17
18
19
20
21

Attachment and uptake of CF

CF was observed uniformly bound to the plasma membrane of soybean protoplasts (Tanchak *et al.* 1984). CF attachment was believed to occur by electrostatic binding and not by attachment to specific receptors. Thus uptake of CF simply reflects normal membrane internalisation. CF uptake by coated pits (Fig. 15) occurred within seconds at room temperature. The process often involved deep pits but never pits with very long necks, as described in some animal cells (e.g. Willingham & Pastan 1983).

Coated pits were released from the plasma membrane to form coated vesicles (Fig. 16). Some coated vesicles seemed to lose their coat rapidly and partially coated vesicles were observed after 2 min. However, other CF-labelled coated vesicles retained their coats and appeared deep within the cytoplasm after 5 min. Serial sectioning analysis confirmed the existence of free labelled coated vesicles in the cytoplasm (Fowke *et al.* 1989). Animal coated vesicles may also exist in the cytoplasm as separate organelles (van Deurs *et al.* 1984).

Attachment and internalisation of CF as well as lectin–colloidal gold conjugates by coated membranes have also been reported for protoplasts of bean leaves and carrot cell cultures (Joachim & Robinson 1984; Hillmer *et al.* 1986). Binding at 0 °C was patchy but considerable label was internalised when samples were warmed.

Appearance of CF in the PCR

Within 2 min of internalisation of CF by coated vesicles in soybean protoplasts, the first CF appeared in membranes of the PCR (Fig. 17) and accumulated with time (Fig. 18, and Tanchak *et al.* 1988). The CF was localised in the tubular elements of the PCR but never in the membrane dilation. Lectin–gold conjugates endocytosed by protoplasts of bean leaf and carrot cells also labelled the PCR but it is not clear how early this labelling occurred after uptake (Hillmer *et al.* 1986). Correlating this with ultrastructural data on CF uptake, recent studies of fluid phase uptake by soybean protoplasts using membrane-impermeant fluorescent probes demonstrated an endosomal compartment, possibly the PCR. The kinetics of probe uptake were similar to those seen in animal cells (Griffing 1988). The fluorescent probe Lucifer Yellow was also shown to accumulate in tiny cytoplasmic particles (PCR?) in carrot protoplasts, although uptake was much slower than in the soybean (Hillmer *et al.* 1989).

The PCR observed in soybean protoplasts appeared to be morphologically similar to a class of animal endosomes known by a number of names including *trans*-reticular Golgi (Willingham & Pastan 1984) and CURL (compartment of uncoupling of receptor and ligand; Geuze *et al.* 1983). These organelles are thought to be involved in receptor–ligand dissociation and/or sorting events. Because of the structural similarities between the PCR and these endosomes and the rapid labelling of the PCR by endocytosed markers in soybean protoplasts, it appears that the PCR functions as an endosome in plant cells. Using a variety of techniques including cell-free systems, it has been possible to distinguish early and late endosomes in animal cells (see Gruenberg & Howell 1989). Early endosomes were located near the plasma membrane and differed in morphology, density, membrane chemistry and pH from the late endosomes, which were usually located in the perinuclear region of the cell. Further studies are necessary to determine similar parameters for the PCR in plants.

Labelling of Golgi with CF

Golgi of soybean protoplasts were often located in the perinuclear region. They consisted of four to eight thin, flattened, stacked cisternae, each cisterna having a slightly dilated rim from which tubules frequently projected into the surrounding cytoplasm. The earliest Golgi labelling occurred within 4 min with CF appearing at the periphery of only one or two cisternae. Labelling of more cisternae occurred after longer exposure to CF, but the peripheral pattern was retained (Fig. 19). Similar results were reported in other protoplasts labelled with CF and lectin–gold conjugates (Joachim & Robinson 1984; Hillmer *et al.* 1986) and in root cells labelled with metal salts (Hübner *et al.* 1985). The movement of label from the plasma membrane to Golgi has important implications, particularly relating to membrane recycling in plants (see below).

The major transport during receptor-mediated endocytosis in animals occurs from the plasma membrane to lysosomes via endosomes. There are many reports of surface labels moving to the reticular elements on the *trans* side of Golgi but only a few cases of delivery of molecules into Golgi cisternae. One example is the transferrin receptor, which is a plasma-membrane-bound protein and is transported into Golgi cisternae of cultured myeloma cells (Woods *et al.* 1986).

Labelling of MVBs with CF

Labelling of MVBs in soybean protoplasts was first detected after 6 min of exposure to CF but only one or two CF particles were observed in any one section of a MVB. By 12 min the labelling of MVBs had increased significantly (Fig. 20) and with longer labelling times MVBs continued to accumulate CF (Fig. 21; and Tanchak *et al.* 1984; Tanchak 1987; Tanchak & Fowke 1987; Record & Griffing 1988). The CF particles were localised on the inner surface of the limiting membrane and on the outer surface of the internal vesicles. Hillmer *et al.* (1986) also demonstrated labelling of MVBs in protoplasts of bean leaf and carrot cells with lectin–gold conjugates, but they did not establish the time course of labelling.

The origin of MVBs in soybean protoplasts is unclear. However, the time course of labelling suggests that MVBs receive their CF label from the PCR and/or Golgi, since these organelles are labelled prior to the MVBs. The major traffic to the MVB is believed to be via the PCR rather than the Golgi (see discussion below). Two mechanisms of transfer from PCR to MVB are possible (Fig. 22). First, the dilation of the PCR, which contains vesicles, could separate from the tubules of the PCR and mature into a MVB. This option seems unlikely, since the MVB is heavily labelled with CF but the PCR dilation never contains CF. Alternatively the PCR and MVB may represent two separate compartments within the cell, perhaps analogous to early and late endosomes in animal cells (Gruenberg & Howell 1989), and transfer between them may be mediated by vesicular transport. The presence of coated pits and possible remnants of coated pits (plaques) on the surface of MVBs suggests that coated vesicles might be involved in this type of transport. Transfer of CF from Golgi to MVBs might also occur via coated vesicles. The presence of degradative enzymes in coated vesicles near Golgi and in MVBs (see e.g. Griffing & Fowke 1985) is consistent with this suggestion, since these enzymes are processed in Golgi (see Kornfeld & Mellman 1989). If a coated vesicle containing CF bound to its inner membrane surface fused with a MVB, the CF would end up on the inner surface of the MVB limiting membrane, as observed. Subsequent invagination of this area of membrane (see e.g. Fig. 12) would produce an internal vesicle with CF on its outer surface, also as observed. Formation of vesicles within MVBs by such an invagination mechanism has been suggested for both plants (see e.g. Halperin & Jensen 1967) and animals (see e.g. Smith & Farquhar 1966).

It is interesting that CF presented to zoospores of the fungus *Aphanomyces euteiches* (Cerenius *et al.* 1988) was internalised via coated membranes and finally deposited in MVBs similar to those described in plant and animal cells. MVBs of some fungi are known to contain hydrolytic enzymes (Kazama 1973).

MVBs ultimately fuse with lysosomes in animal cells (Gruenberg & Howell 1989; Kornfeld & Mellman 1989) but fuse with vacuoles in plant

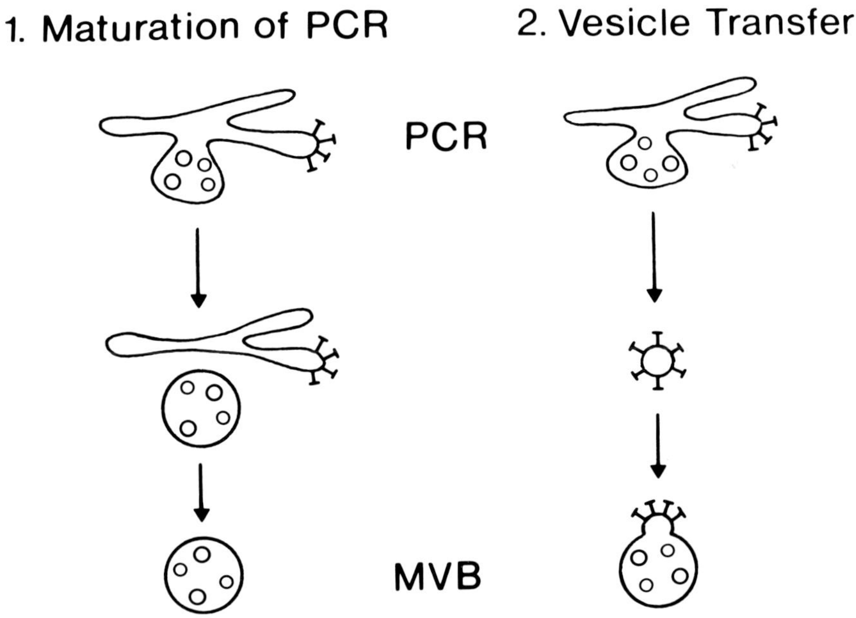

Fig. 22. Diagram illustrating two possible mechanisms for the transfer of CF from PCR to MVB. The results from CF labelling of soybean protoplasts favour vesicle transfer.

cells (see below). In one specialised type of animal cell, the reticulocyte, MVBs fused with the plasma membrane releasing their contents including their internal vesicles to the external medium (Harding *et al.* 1983). The possibility that some MVBs from soybean protoplasts may behave in a similar manner cannot be ruled out. Such a fusion event would be very difficult to record using protoplasts. However, the presence of plasma membrane invaginations containing small vesicles (plasmalemmasomes) in intact soybean cells (Tanchak 1987) is consistent with this idea. Further work is necessary to evaluate this possibility.

Transfer of CF to the vacuole

Soybean protoplasts exposed to CF for periods of at least 1 h contained clusters of CF in the central vacuole (Fig. 23 (a) and (b)) as well as in some of the peripheral vacuoles (Tanchak & Fowke 1987; Record & Griffing 1988). The CF particles were often associated with groups of small vesicles similar in appearance and size to those observed in MVBs in the cytoplasm. This observation gave rise to the suggestion that the CF

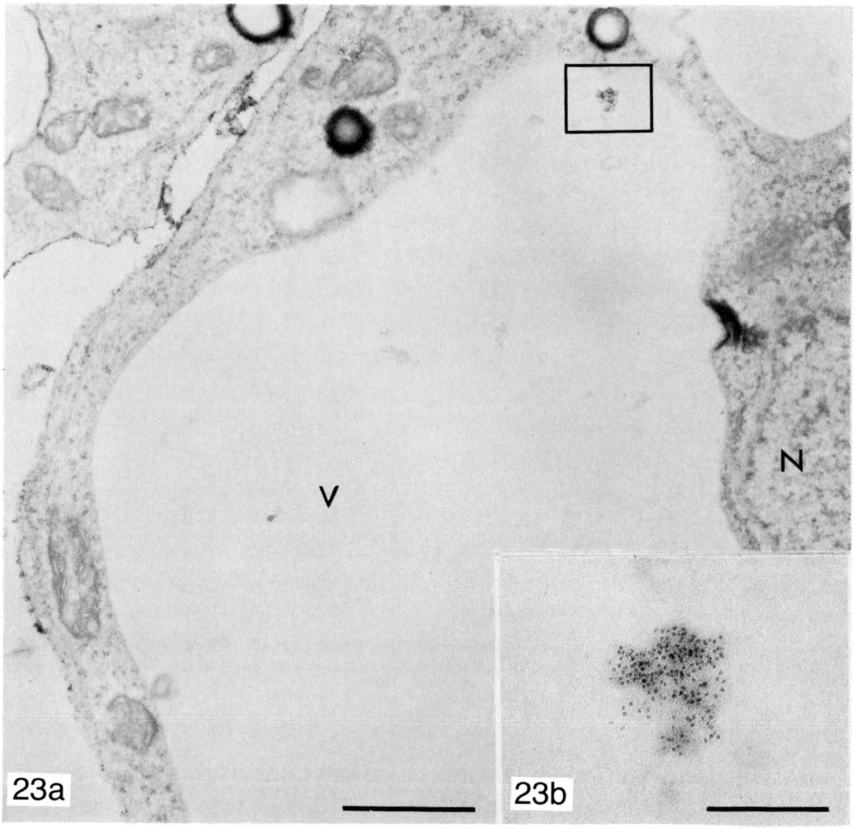

Fig. 23 (a). Low magnification view of a soybean protoplast exposed to CF for 60 min. The cluster of CF particles in the central vacuole is outlined in black and is enlarged in (b). V, vacuole; N, nucleus. The bar represents 1 μm. (From Tanchak & Fowke 1987.) (b) Enlargement from (a) showing cluster of CF particles located in the central vacuole. The bar represents 200 nm.

was transferred to the vacuole by fusion of MVBs with the vacuole tonoplast (Tanchak & Fowke 1987). Similarly in animal cells, internalised molecules were delivered from MVBs (late endosomes) to lysosomes by a fusion process (Gruenberg & Howell 1989).

It should be noted that the results of Joachim & Robinson (1984) differ from the results of the CF uptake studies using soybean protoplasts. In their ultrastructural studies of endocytosis by bean leaf and carrot cell protoplasts, they observed labelling of only small vacuoles and never of the central vacuole (see e.g. Joachim & Robinson 1984; Hillmer *et al.* 1986). Furthermore, they were unable to label the central vacuole of carrot protoplasts with fluorescent Lucifer Yellow even after prolonged exposure (Hillmer *et al.* 1989). These observations are difficult to interpret in the light of recent light microscope studies by other groups. Wright & Oparka (1989) exposed protoplasts from a variety of plant species to fluorescent Lucifer Yellow and demonstrated accumulation of the dye in the vacuoles of all types of protoplast. Similarly, Griffing (1988) observed movement of externally supplied fluorescent molecules to the vacuoles of soybean protoplasts. The evidence presented in both of these studies suggests that fluid phase endocytosis does occur. Recent studies with intact yeast (Reizman 1985) and plant (Oparka *et al.* 1988; Hillmer *et al.* 1989; Horn *et al.* 1989) cells exposed to externally supplied fluorescent molecules also showed labelling of the vacuoles.

In summary, the results obtained from ultrastructural studies of CF uptake by soybean protoplasts demonstrate an endocytotic pathway in plants. This pathway, illustrated diagrammatically in Fig. 24, involves endocytosis by coated pits and sequential delivery to coated vesicles, smooth vesicles, the PCR and Golgi, MVBs and finally to the cell vacuole.

Possible roles for the endocytotic pathway

The CF uptake data may in fact reveal two distinct routes of internalisation in plant cells: (1) plasma membrane to Golgi for membrane recycling and repair and (2) plasma membrane to vacuole for degradative processes.

Plasma membrane to Golgi

Membrane recycling

The ultrastructural evidence indicates clearly that regions of the plasma membrane with attached CF are transferred to Golgi in plant protoplasts. This transfer may reflect a major mechanism for recycling of plasma membrane to Golgi to compensate for their secretory activity. The growth and development of cell walls on plant cells involves constant secretion of matrix polysaccharides by smooth Golgi vesicles and this process adds a significant amount of membrane to the plasma membrane (for a review, see Robinson & Kristen 1982). It is clear that plant cells rapidly replace their plasma membrane by internalisation and the return route into the cell appears to involve coated pits and coated vesicles (for reviews, see Steer 1988; Steer & O'Driscoll, this volume). Coated pits

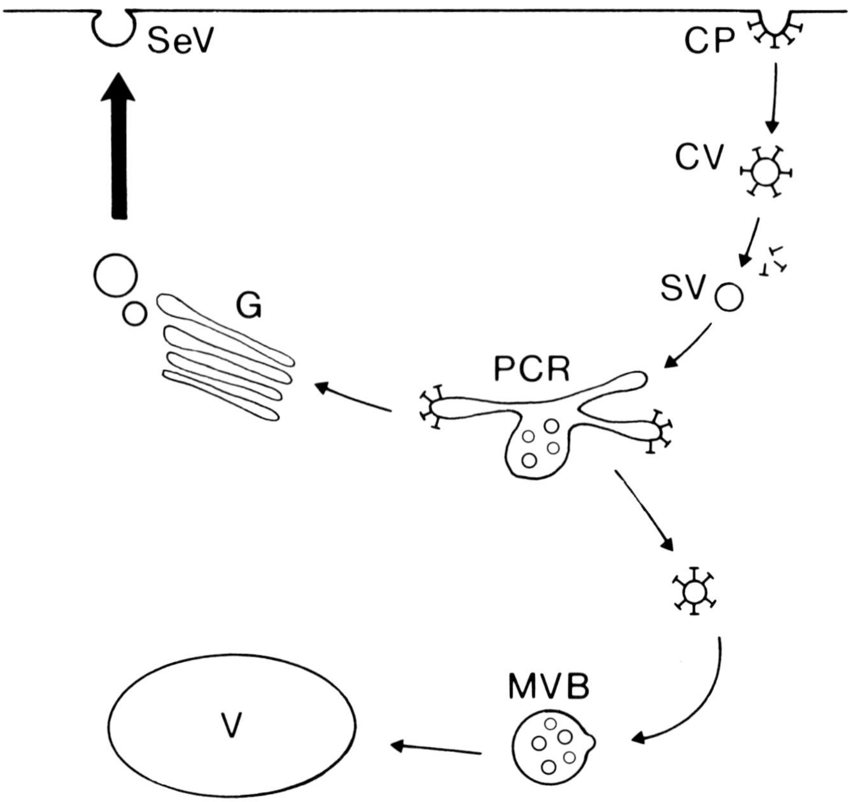

Fig. 24. Diagrammatic representation of endocytotic pathway in soybean protoplasts. CF is endocytosed by coated pits (CP) to coated vesicles (CV), which uncoat to form smooth vesicles (SV). CF is then delivered to the PCR and Golgi (G). This pathway may retrieve plasma membrane to compensate for excess plasma membrane contributed by the deposition of cell wall matrix components (heavy arrow) by smooth Golgi secretion vesicles (SeV). Some CF in the PCR is also transferred to MVBs, possibly by coated vesicles (see Fig. 22), and is finally delivered to the cell vacuole (V). This pathway may provide a mechanism for removing foreign molecules from the cell surface for degradation.

and coated vesicles are most frequently observed in actively growing plant cells. Emons & Traas (1986) calculated that growing cells could recycle their complete plasma membrane in 20–40 min and suggested that coated membranes were responsible for this process.

Studies of wounding of algal cells (O'Neil & La Claire 1984; Foissner

1988) also support this proposed role for coated pits and coated vesicles in membrane recycling. O'Neil & La Claire (1984) observed a direct correlation between the appearance of numerous large coated pits and the dramatic reduction of plasma membrane surface area during the recovery of wounded *Boergesenia forbesii* cells.

High frequencies of coated pits and coated vesicles have been reported during other processes involving rapid plasma membrane and cell wall deposition but the coated membranes were interpreted as exocytotic. For example, Franke & Herth (1974) estimated that 60% of the plasma membrane on rapidly forming cell plates of dividing *Happlopappus gracilis* cells was coated and they proposed that cytoplasmic coated vesicles were transferred to the developing cell plate. Also Robertson & Lyttleton (1982) suggested that the numerous coated membranes associated with rapidly growing infection threads during rhizobial infection of white clover root hairs were derived from Golgi in the cytoplasm. In both of these cases, retrieval of the plasma membrane by endocytosis would be consistent with the observations.

While direct membrane recycling from the plasma membrane to Golgi is certainly possible, there are reasons for arguing that the plasma membrane-derived vesicles fuse with another organelle, probably the PCR, prior to transfer of membrane to the Golgi. Endocytosis in soybean protoplasts must involve more than mere membrane recycling, since MVBs and vacuoles, the lytic compartments of the protoplasts, also become labelled. Thus a significant portion of the membrane material and/or lumenal contents internalised via endocytosis is destined to be degraded rather than recycled. It seems unlikely that two distinct populations of coated pits exist on the plasma membrane with one population involved in the recycling of plasma membrane to the Golgi and a second population involved in the delivery of material to the lytic compartment. At least this does not appear to be the case in animal cells; different types of receptor with different fates following internalisation via endocytosis occur randomly at the cell surface and can be found within the same coated pits and coated vesicles (Geuze *et al.* 1984). Furthermore, the sorting of different receptor–ligand complexes from one another and the sorting of the components of dissociated receptor–ligand complexes occurs intracellularly within endosomes (Gruenberg & Howell 1989). Therefore, it seems probable that, in plant cells, membrane constituents destined to be recycled and those to be degraded are localised within the same coated pit and that sorting occurs sometime after internalisation. The PCR seems the most likely organelle where sorting events could occur within soybean protoplasts. Constituents to be recycled would then be transferred from the PCR to Golgi.

Plasma membrane repair

The Golgi of both animal and plant cells have been shown to contain a number of glycosyltransferases which are involved in the post-translational modification of newly synthesised proteins (for reviews, see Farquhar 1985; Robinson & Kristen 1982). It is possible that these enzymes also function to repair damaged oligosaccharide chains on recycling glycoproteins. For example, partial resialyation of asialotransferrin type 3 occurred during diacytosis by rat hepatocytes (Regoeczi *et al.* 1982). In addition, it has been demonstrated that in K 562 (animal) cells the cell surface transferrin receptors which had been asialyated by neuraminadase treatment were resialyated following internalisation (Snider & Rogers 1985). It is thought that these repair processes are mediated by the Golgi (Regoeczi *et al.* 1982; Snider & Rogers 1985) and it is possible that plant Golgi operate in a similar manner. Therefore, the labelling of the Golgi with CF may indicate the occurrence of a pathway for repair of damaged membrane constituents in addition to a possible membrane recycling pathway.

Plasma membrane to vacuole

Receptor-mediated endocytosis

The striking similarity between the pathway observed in soybean protoplasts and pathways for receptor-mediated endocytosis in many animal cells raises the interesting possibility that plant cells also are involved in a highly specific process of endocytosis. Likely candidates for this type of specific internalisation include plant growth regulators, fungal toxins and elicitors. Plant growth substances are sufficiently small and hydrophilic to be able to pass directly through the plasma membrane by diffusion. Elicitors and toxins, however, are too large and/or hydrophobic to pass easily through the plasma membrane. Both types of macromolecule are produced by plant pathogens and have rapid effects on the host plant cells.

Many plant cells produce phytoalexins when stimulated by elicitors as part of the defence response. The production of phytoalexins is thought to prevent growth of the invading microorganism at the site of attempted infection. Soybean tissues produce and accumulate phytoalexins in response to the fungus *Phytophthora megasperma* or to a β-glucan elicitor isolated from the fungal cell walls (Ebel & Grisebach 1988). High affinity binding sites for glucan elicitor have been described in soybean plasma membrane-enriched fractions (Schmidt & Ebel 1987) and the elicitor will bind specifically to the plasma membrane of soybean protoplasts *in vivo*

(Cosio *et al.* 1988). The study of Horn *et al.* (1989) suggests that labelled glucan elicitors can penetrate the cell walls and bind to the surface of cultured soybean cells, where they are internalised by a temperature- and energy-dependent process and are then transferred to the cell vacuole. Ultrastructural studies are now required to determine whether this uptake is in fact mediated by coated pits and coated vesicles.

Plant cells are also exposed to different fungal toxins and evidence suggests that they bind to specific sites on the plasma membrane. Fusicoccin, the major toxin produced by *Fusicoccum amygdali* is known to bind to plasma membrane receptors and evoke a variety of symptoms in plant cells (Marre 1979; Aducci *et al.* 1982). Future experiments might examine the possibility that the endocytotic pathway described above operates to remove toxins from the cell surface via coated vesicles and transfer them to the vacuole for degradation.

Other possible roles for coated vesicles in plants

Research with animal cells has shown that, in addition to functioning in receptor-mediated endocytosis, coated vesicles are involved in the process of protein sorting and transport. Lysosomal enzymes, for example, are manufactured on the endoplasmic reticulum, transferred to the Golgi and eventually moved to lysosomes via coated vesicles (see Farquhar 1985).

Plant coated vesicles may also be involved in protein sorting and transport. Ultrastructural studies indicate that plant cells contain two populations of coated vesicles, those near the plasma membrane, which are believed to result from endocytosis, as well as smaller coated vesicles further from the plasma membrane and often localised near Golgi (van der Valk & Fowke 1981; Coleman *et al.* 1987). Using cytochemistry, Griffing & Fowke (1985) localised the enzyme peroxidase in the latter population of coated vesicles. Recent biochemical studies of pea cotyledons indicate that coated vesicles are also involved in intracellular transport of storage proteins to protein bodies during seed development (Harley & Beevers 1989; Robinson *et al.* 1989).

Available evidence does not directly support a role for coated vesicles in exocytosis in plant cells but such a possibility has not been ruled out. Indeed, results of many earlier studies of coated vesicles in cells rapidly depositing cell walls were interpreted to indicate exocytosis of wall precursors (for a discussion, see Fowke *et al.* 1983*a*). Cytochemical studies have demonstrated polysaccharides in plant coated vesicles (Ryser 1979; Hawes *et al.* 1989) but the direction of movement of the vesicles was not established. The enzyme glucan synthase I, which produces a β-1-4-linked glucan, was enriched in coated vesicle fractions isolated from soybean protoplasts (Griffing *et al.* 1986). This enzyme could represent the cellulose synthase proenzyme (Hayashi & Matsuda 1981), which may be transported to the plasma membrane via coated vesicles. Rosettes of particles which are believed to be involved in the process of cellulose synthesis have been located in plant Golgi and are thought to be delivered to the plasma membrane via Golgi-derived vesicles (Haigler & Brown 1986). However, these vesicles did not appear to be coated. Further work is necessary to determine whether coated vesicles do play a role in exocytosis in plants.

Conclusions

Ultrastructural studies have clarified the distribution of coated pits and coated vesicles in a variety of plant cells and protoplasts and have demonstrated a direct correlation between their numbers and cell growth. The electron microscope has also provided important information regarding the basic morphology of these structures and other organelles of the endocytotic pathway in plants and has shown that they closely resemble their counterparts in animals.

Over the past six years electron microscopy has also played a key role in illustrating the uptake and transport of non-specific labels by plant protoplasts and cells. It is clear from endocytosis studies using soybean protoplasts that CF can be moved sequentially from coated pits in the plasma membrane via coated vesicles to the PCR and Golgi, MVBs and finally to the vacuole. This transport may reflect two major pathways from the cell surface, first to the Golgi for plasma membrane recycling and repair, and second to the major lytic compartment of the cell, the vacuole, for degradative processes. The remarkable similarity between this process in plants and receptor-mediated endocytosis in animals suggests that plant cells may also internalise macromolecules in a very specific fashion. Likely candidates for such uptake include plant growth substances, fungal elicitors and toxins. Research is now required to determine which molecules utilise the pathway, identify specific receptors on the plasma membrane and isolate and characterise the organelles involved in the endocytotic pathway. The next five years should provide exciting information regarding endocytosis in plant cells.

Acknowledgements

The contributions of numerous scientists and research associates working in this laboratory are gratefully acknowledged. A special thanks goes to

Pat Rennie and Dennis Dyck for help in preparing this manuscript. The research work from this laboratory has been generously supported by the Natural Sciences and Engineering Research Council of Canada.

References

Aducci, P., Ballilo, A., Federico, R. & Montesano, L. (1982). Studies on fusicoccin-binding sites. In *Plant Growth Substances 1982*, ed. P.F. Wareing, pp. 394–404. London: Academic Press.

Boller, T. & Kende, H. (1979). Hydrolytic enzymes in the central vacuole of plant cells. *Plant Physiology*, **63**, 1123–32.

Buvat, R. & Robert, G. (1979). Vacuole formation in the actively growing root meristem of barley (*Hordeum sativum*). *American Journal of Botany*, **66**, 1219–37.

Cerenius, L., Rennie, P.J. & Fowke, L.C. (1988). Endocytosis of cationized ferritin by zoospores of the fungus *Aphanomyces euteiches*. *Protoplasma*, **144**, 119–24.

Coleman, J., Evans, D. & Hawes, C. (1988). Plant coated vesicles. *Plant, Cell and Environment*, **11**, 669–84.

Coleman, J., Evans, D., Hawes, C., Horsley, D. & Cole, L. (1987). Structure and molecular organization of higher plant coated vesicles. *Journal of Cell Science*, **88**, 35–45.

Cosio, E., Popperyl, H., Schmidt, W.E. & Ebel, J. (1988). High affinity binding of fungal β-glucan fragments to soybean (*Glycine max* L.) microsomal fractions and protoplasts. *European Journal of Biochemistry*, **175**, 309–15.

Cram, W.J. (1980). Pinocytosis in plants. *New Phytologist*, **84**, 1–17.

Ebel, J. & Grisebach, H. (1988). Defense strategies of soybean against the fungus *Phytophthora megasperma* f. sp. *glycinea*. A molecular analysis. *Trends in Biochemical Sciences*, **13**, 23–7.

Emons, A.M.C. & Traas, J.A. (1986). Coated pits and coated vesicles on the plasma membrane of plant cells. *European Journal of Cell Biology*, **41**, 57–64.

Farquhar, M.G. (1985). Progress in unravelling pathways of Golgi traffic. *Annual Review of Cell Biology*, **1**, 447–88.

Fine, R.E. & Ockleford, C.D. (1984). Supramolecular cytology of coated vesicles. *International Review of Cytology*, **91**, 1–43.

Foissner, I. (1988). The relationship of echinate inclusions and coated vesicles on wound healing in *Nitella flexilis* (Characeae). *Protoplasma*, **142**, 164–75.

Fowke, L.C. (1986). Ultrastructural cytology of cultured plant tissues, cells and protoplasts. In *Cell Culture and Somatic Cell Genetics of Plants*, vol. 3, ed. I.K. Vasil, pp. 323–42. New York: Academic Press.

Fowke, L.C. & Gamborg, O.L. (1980). Applications of protoplasts to the study of plant cells. *International Review of Cytology*, **68**, 9–51.

Fowke, L.C., Griffing, L.R., Mersey, B.G. & van der Valk, P. (1983*a*). Protoplasts for studies of the plasma membrane and associated cell organelles. In *Protoplasts 1983, Lecture Proceedings*, ed. I. Potrykus, C.T. Harms, A. Hinnen, R. Hutter, P.J. King & R.D. Shillito, pp. 101–10. Basel: Birkhauser.

Fowke, L.C. & Pickett-Heaps, J.D. (1978). Electron microscope study of vegetative division in two species of *Marchantia*. *Canadian Journal of Botany*, **56**, 467–75.

Fowke, L.C., Rennie, P.J. & Constabel, F. (1983*b*). Organelles associated with the plasma membrane of tobacco leaf protoplasts. *Plant Cell Reports*, **2**, 292–5.

Fowke, L.C., Tanchak, M.A. & Rennie, P.J. (1989). Serial section analysis of coated pits and coated vesicles in soybean protoplasts. *Cell Biology International Reports*, **13**, 419–25.

Franke, W.W. & Herth, W. (1974). Morphological evidence for *de novo* formation of plasma membrane from coated vesicles in exponentially growing cultured plant cells. *Experimental Cell Research*, **89**, 447–51.

Geuze, H.J., Slot, J.W., Strous, G.J.A.M., Lodish, H.F. & Schwartz, A.L. (1983). Intracellular site of asialoglycoprotein receptor–ligand uncoupling: double-label immunoelectron microscopy during receptor-mediated endocytosis. *Cell*, **32**, 277–87.

Geuze, H.J., Slot, J.W., Strous, G.J.A.M., Peppard, J., von Figura, K., Hasilik, K. & Schwartz, A.L. (1984). Intracellular receptor sorting during endocytosis: comparative immunoelectron microscopy of multiple receptors in rat liver. *Cell*, **37**, 195–204.

Goldstein, J.L., Brown, M.S., Anderson, R.G.W., Russell, D.W. & Schneider, W.J. (1985). Receptor-mediated endocytosis: concepts emerging from the LDL receptor system. *Annual Review of Cell Biology*, **1**, 1–39.

Gradman, D. & Robinson, D.G. (1989). Does turgor prevent endocytosis in plant cells? *Plant, Cell and Environment*, **12**, 151–4.

Griffing, L.R. (1988). Fluid-phase and membrane-bound transport to the endocytic compartment in plants. *Current Topics in Plant Biochemistry and Physiology*, **7**, 101–11.

Griffing, L.R. & Fowke, L.C. (1985). Cytochemical localization of peroxidase in soybean suspension culture cells and protoplasts: intracellular vacuole differentiation and presence of peroxidase in coated vesicles and multivesicular bodies. *Protoplasma*, **128**, 22–30.

Griffing, L.R., Mersey, B.G. & Fowke, L.C. (1986). Cell-fractionation analysis of glucan synthase I and II distribution and polysaccharide secretion in soybean protoplasts. *Planta*, **167**, 175–82.

Gruenberg, J. & Howell, K.E. (1989). Membrane traffic in endocytosis: insights from cell-free assays. *Annual Review of Cell Biology*, **5**, 453–82.

Guilliermond, A. (1941). *The Cytoplasm of the Plant Cell*. Waltham, MA: Chronica Botanica Company.

Gunning, B. E. S. & Steer, M. W. (1975). *Ultrastructure and the Biology of Plant Cells*. London: Arnold.

Haigler, C. H. & Brown, R. M. (1986). Transport of rosettes from the Golgi apparatus to the plasma membrane in isolated mesophyll cells of *Zinnia elegans* during differentiation to tracheary elements in suspension culture. *Protoplasma*, **134**, 111–20.

Hakman, I., Rennie, P. & Fowke, L. C. (1987). A light and electron microscope study of *Picea glauca* (White spruce) somatic embryos. *Protoplasma*, **140**, 100–9.

Halperin, W. (1969). Ultrastructural localization of acid phosphatase in cultured cells of *Daucus carota*. *Planta*, **88**, 91–102.

Halperin, W. & Jensen, W. A. (1967). Ultrastructural changes during growth and embryogenesis in carrot cell cultures. *Journal of Ultrastructure Research*, **18**, 428–43.

Harding, C., Heuser, J. & Stahl, P. (1983). Receptor-mediated endocytosis of transferrin and recycling of the transferrin receptor in rat reticulocytes. *Journal of Cell Biology*, **97**, 329–39.

Harley, S. M. & Beevers, L. (1989). Coated vesicles are involved in the transport of storage proteins during seed development in *Pisum sativum* L. *Plant Physiology*, **91**, 674–8.

Hawes, C., Coleman, J., Evans, D. & Cole, L. (1989). Recent advances in the study of plant coated vesicles. *Cell Biology International Reports*, **13**, 119–28.

Hawes, C. & Martin, M. (1986). Deep etching of plant cells: cytoskeleton and coated pits. *Cell Biology International Reports*, **10**, 985–7.

Hayashi, T. & Matsuda, K. (1981). Biosynthesis of xyloglucan in suspension-cultured soybean cells. Evidence that the enzyme system of xyloglucan synthesis does not contain β-1,4-glucan 4-β-D-glucosyltransferase activity (E. C. 2.4.1.12). *Plant and Cell Physiology*, **22**, 1571–84.

Hillmer, S., Depta, H. & Robinson, D. G. (1986). Confirmation of endocytosis in higher plant protoplasts using lectin–gold conjugates. *European Journal of Cell Biology*, **41**, 142–9.

Hillmer, S., Freundt, H. & Robinson, D. G. (1988). The partially coated reticulum and its relationship to the Golgi apparatus in higher plant cells. *European Journal of Cell Biology*, **47**, 206–12.

Hillmer, S., Quader, H., Robert-Nicoud, M. & Robinson, D. G. (1989). Lucifer Yellow uptake in cells and protoplasts of *Daucus carota* visualized by laser scanning microscopy. *Journal of Experimental Botany*, **40**, 417–23.

Hippe, S., During, K. & Kreuzaler, F. (1989). *In situ* localization of a foreign protein in transgenic plants by immunoelectron microscopy following high pressure freezing. Freeze substitution and low temperature embedding. *European Journal of Cell Biology*, **50**, 230–4.

Horn, M. A., Heinstein, P. F. & Low, P. S. (1989). Receptor-mediated endocytosis in plant cells. *Plant Cell*, **1**, 1003–9.

Hübner, R., Depta, H. & Robinson, D. G. (1985). Endocytosis in maize root cap cells. Evidence obtained using heavy metal salt solutions. *Protoplasma*, **129**, 214–22.

Joachim, S. & Robinson, D. G. (1984). Endocytosis of cationic ferritin by bean leaf protoplasts. *European Journal of Cell Biology*, **34**, 212–16.

Kazama, F. (1973). Ultrastructure of *Thraustochytrium* sp. zoospores. III. Cytolysosomes and acid phosphatase distribution. *Archives of Microbiology*, **89**, 95–104.

Kornfeld, S. & Mellman, I. (1989). The biogenesis of lysosomes. *Annual Review of Cell Biology*, **5**, 483–525.

Marchant, R. & Robards, A. W. (1968). Membrane systems associated with the plasmalemma of plant cells. *Annals of Botany*, **32**, 457–71.

Marre, E. (1979). Fusicoccin, a tool in plant physiology. *Annual Review of Plant Physiology*, **30**, 273–88.

Matile, P. (1978). Biochemistry and function of vacuoles. *Annual Review of Plant Physiology*, **29**, 193–213.

Nakamura, S. & Miki-Hirosige, H. (1982). Coated vesicles and cell plate formation in the microspore mother cell. *Journal of Ultrastructure Research*, **80**, 302–11.

Newcomb, E. H. (1980). Coated vesicles: their occurrence in different plant cell types. In *Coated Vesicles*, ed. C. Y. Ockleford & A. Whyte, pp. 55–68. Cambridge: Cambridge University Press.

Nishimura, M. & Beevers, H. (1979). Hydrolysis of protein in vacuoles isolated from higher plant tissue. *Nature (London)*, **277**, 412–13.

Ockleford, C. D. & Whyte, A. (1980). *Coated Vesicles*. Cambridge: Cambridge University Press.

O'Neil, R. M. & La Claire, J. W. (1984). Mechanical wounding induces the formation of extensive coated membranes in giant algal cells. *Science*, **225**, 331–3.

Oparka, K. J., Robinson, D., Prior, D. A. M., Derrick, P. & Wright, K. M. (1988). Uptake of lucifer yellow CH into intact barley roots: evidence for fluid-phase endocytosis. *Planta*, **176**, 541–7.

Pearse, B. M. & Bretscher, M. S. (1981). Membrane recycling by coated vesicles. *Annual Review of Biochemistry*, **50**, 85–101.

Pesacreta, T. C. & Lucas, W. J. (1984). Plasma membrane coat and a coated vesicle-associated reticulum of membranes: their structure and possible interrelationship in *Chara corallina*. *Journal of Cell Biology*, **98**, 1537–45.

Pesacreta, T. C. & Lucas, W. J. (1985). Presence of a partially-coated reticulum in angiosperms. *Protoplasma*, **125**, 173–84.

Record, R. D. & Griffing, L. R. (1988). Convergence of the endocytic and lysosomal pathways in soybean protoplasts. *Planta*, **176**, 425–32.

Regoeczi, E., Chindemi, P. A., Debanne, M. T. & Charlwood, P. A.

(1982). Partial resialylation of human asialotransferrin type 3 in the rat. *Proceedings of the National Academy of Sciences, USA*, **79**, 2226–30.

Reizman, H. (1985). Endocytosis in yeast: several of the yeast secretory mutants are defective in endocytosis. *Cell*, **40**, 1001–9.

Robertson, J.G. & Lyttleton, P. (1982). Coated and smooth vesicles in the biogenesis of cell walls, plasma membranes, infection threads and peribacteroid membranes in root hairs and nodules of white clover. *Journal of Cell Science*, **58**, 63–78.

Robinson, D.G., Balusek, K. & Freundt, H. (1989). Legumin antibodies recognize polypeptides in coated vesicles isolated from developing pea cotyledons. *Protoplasma*, **150**, 79–82.

Robinson, D.G. & Kristen, U. (1982). Membrane flow via the Golgi apparatus of higher plant cells. *International Review of Cytology*, **77**, 89–127.

Ryser, U. (1979). Cotton fibre differentiation: occurrence and distribution of coated vesicles and smooth vesicles during primary and secondary wall formation. *Protoplasma*, **98**, 223–39.

Saxton, M.J. & Breidenbach, R.W. (1988). Receptor-mediated endocytosis in plants is energetically possible. *Plant Physiology*, **86**, 993–5.

Schmidt, W.E. & Ebel, J. (1987). Specific binding of a fungal glucan phytoalexin elicitor to membrane fractions from soybean *Glycine max*. *Proceedings of the National Academy of Sciences, USA*, **84**, 4117–121.

Smith, R.E. & Farquhar, M.G. (1966). Lysosome function in the regulation of the secretory process in cells of the anterior pituitary gland. *Journal of Cell Biology*, **31**, 319–47.

Snider, M.D. & Rogers, O.C. (1985). Intracellular movement of cell surface receptors after endocytosis: resialylation of asialo-transferrin receptor in human erythroleukemia cells. *Journal of Cell Biology*, **100**, 826–34.

Steer, M.W. (1988). Plasma membrane turnover in plant cells. *Journal of Experimental Botany*, **39**, 987–96.

Tanchak, M.A. (1987). Endocytosis of cationized ferritin by soybean (*Glycine max* L.) protoplasts. Ph.D. thesis, University of Saksatchewan.

Tanchak, M.A. & Fowke, L.C. (1987). The morphology of multivesicular bodies in soybean protoplasts and their role in endocytosis. *Protoplasma*, **138**, 173–82.

Tanchak, M.A., Griffing, L.R., Mersey, B.G. & Fowke, L.C. (1984). Endocytosis of cationized ferritin by coated vesicles of soybean protoplasts. *Planta*, **162**, 481–6.

Tanchak, M.A., Rennie, P.J. & Fowke, L.C. (1988). Ultrastructure of the partially coated reticulum and dictyosomes during endocytosis by soybean protoplasts. *Planta*, **175**, 433–41.

Traas, J.A. (1984). Visualization of the membrane bound cytoskeleton and coated pits of plant cells by means of dry cleaving. *Protoplasma*, **119**, 212–18.

van der Valk, P. & Fowke, L.C. (1981). Ultrastructural aspects of coated vesicles in tobacco protoplasts. *Canadian Journal of Botany*, **59**, 1307–13.

van der Valk, P., Rennie, P.J., Connolly, J.A. & Fowke, L.C. (1980). Distribution of cortical microtubules in tobacco protoplasts. An immunofluorescence microscopic and ultrastructural study. *Protoplasma*, **105**, 27–43.

van Deurs, B., Peterson, O.W. & Bundegaard, M. (1984). Identification of free coated pinocytic vesicles in Swiss 3T3 cells. *EMBO Journal*, **3**, 1959–64.

Wileman, T., Harding, C. & Stahl, P. (1985). Receptor-mediated endocytosis. *Biochemical Journal*, **232**, 1–14.

Willingham, M.C. & Pastan, I. (1983). Formation of receptosomes from plasma membrane coated pits during endocytosis: analysis by serial sections with improved membrane labelling and preservation techniques. *Proceedings of the National Academy of Sciences, USA*, **80**, 5617–21.

Willingham, M.C. & Pastan, I. (1984). Endocytosis and exocytosis: current concepts of vesicle traffic in animal cells. *International Review of Cytology*, **92**, 51–92.

Woods, J.W., Doriaux, M. & Farquhar, M.G. (1986). Transferrin receptors recycle to cis and middle as well as trans Golgi cisternae in Ig-secreting myeloma cells. *Journal of Cell Biology*, **103**, 277–86.

Wright, K.M. & Oparka, K.J. (1989). Uptake of Lucifer Yellow CH into plant-cell protoplasts: a quantitative assessment of fluid-phase endocytosis. *Planta*, **179**, 257–64.

J.O.D. COLEMAN, D.E. EVANS, D. HORSLEY AND C.R. HAWES

The molecular structure of plant clathrin and coated vesicles

The transport of materials within eukaryotic cells is central to the maintenance and regulation of cell function, and the traffic of many different molecules can be directed simultaneously. Many membrane-bound and secretory proteins are packaged for translocation by vesicles and transferred to specific organelles or compartments by fusion of the vesicle with acceptor membranes. The process by which certain membrane proteins (i.e. cell surface receptors and their bound ligands) are exclusively recruited into vesicles requires the participation of a characteristic transport agent, the clathrin-coated vesicle. Such vesicles have been shown to initiate a sequence of important transport events in at least two regions of the cell.

1. At the plasma membrane during receptor-mediated endocytosis.
2. At the *trans*-Golgi network during the transfer of newly synthesised proteins to the lysosome.

Coated vesicles were identified in electron micrographs of plant cells soon after they were first described in animal cells (for a review, see Newcomb 1980). Since then the pioneering work of Pearse (1975) has led to tremendous advances in resolving the molecular structure of mammalian coated vesicles and the functions of these organelles. To date, most of what is known on the structure and function of coated vesicles has come from work on mammalian systems. On the other hand, studies on plant coated vesicles were slow to get started, a fact that may be partially attributed to the initial assumption that they were involved solely in exocytotic events.

Although the exact functions of coated vesicles in plant cell have yet to be determined, recent work is shedding light on their involvement in endocytosis and protein transport around the cell (see Fowke *et al.* and Robinson *et al.*, this volume; Harley & Beevers 1989). However, in order to understand their precise mode of action in the capture and transport of macromolecules it will be necessary to characterise, at the molecular level, the components of the coat and their interaction with the vesicle membrane and transported molecules.

In this chapter we focus on the recent advances made in the characterisation of plant clathrin-coated vesicles, summarising information which has been obtained from biochemical and ultrastructural studies including work with specific antibodies. These data are compared with our current knowledge on the molecular structure of yeast and animal clathrin.

Distribution of coated vesicles in plant tissues

Coated vesicles and coated membranes can be considered to be ubiquitous in plant cells. Close examination of many published electron micrographs will reveal that coated vesicles are present in the cytoplasm. These are the typical clathrin-coated vesicles as described in numerous animal cell systems (Ockleford & Whyte 1980). Spiny coated vesicles that are not clathrin coated appear to be restricted to differentiating phloem tissue (Esau & Gill 1970; Newcomb 1980).

Plasma membrane-associated coated vesicles are found in many cell types but appear to be particularly common in cells that are meristematic, growing or are actively involved in transport processes. Cells such as root hairs (Emons & Traas 1986), developing cotton fibres (Ryser 1979), developing pollen (Nakamura & Miki-Hirosige 1982; Sheldon & Dickinson 1983), differentiating vascular tissue (Newcomb 1980), mycorrhizal host cells (Barrosa & Pais 1987) and *Rhizobium*-infected root nodules (Robertson & Lyttleton 1982) are rich in plasma membrane-associated coated vesicles. Coated vesicles associated with the Golgi apparatus and partially coated reticulum have also been reported from a wide range of cell types (see Fowke *et al.*, this volume).

Fig. 1. (a) Transmission electron micrograph of the inside face of the plasma membrane of a dry cleaved carrot suspension-cultured cell (see Hawes 1985). Clathrin-coated pits (arrows) and coated vesicles (double arrows) are prominent features in such preparations. Note also the plasma membrane-associated cortical microtubule (MT) cytoskeleton. The bar represents 100 nm.
(b) Carbon platinum replica of the inside face of an osmotically burst carrot suspension-culture cell protoplast. As in the cells, coated pits are a major component of the membrane face. Under the buffer conditions used, the microtubules are not preserved. The bar represents 100 nm. (From Coleman *et al.* 1987, with permission.)

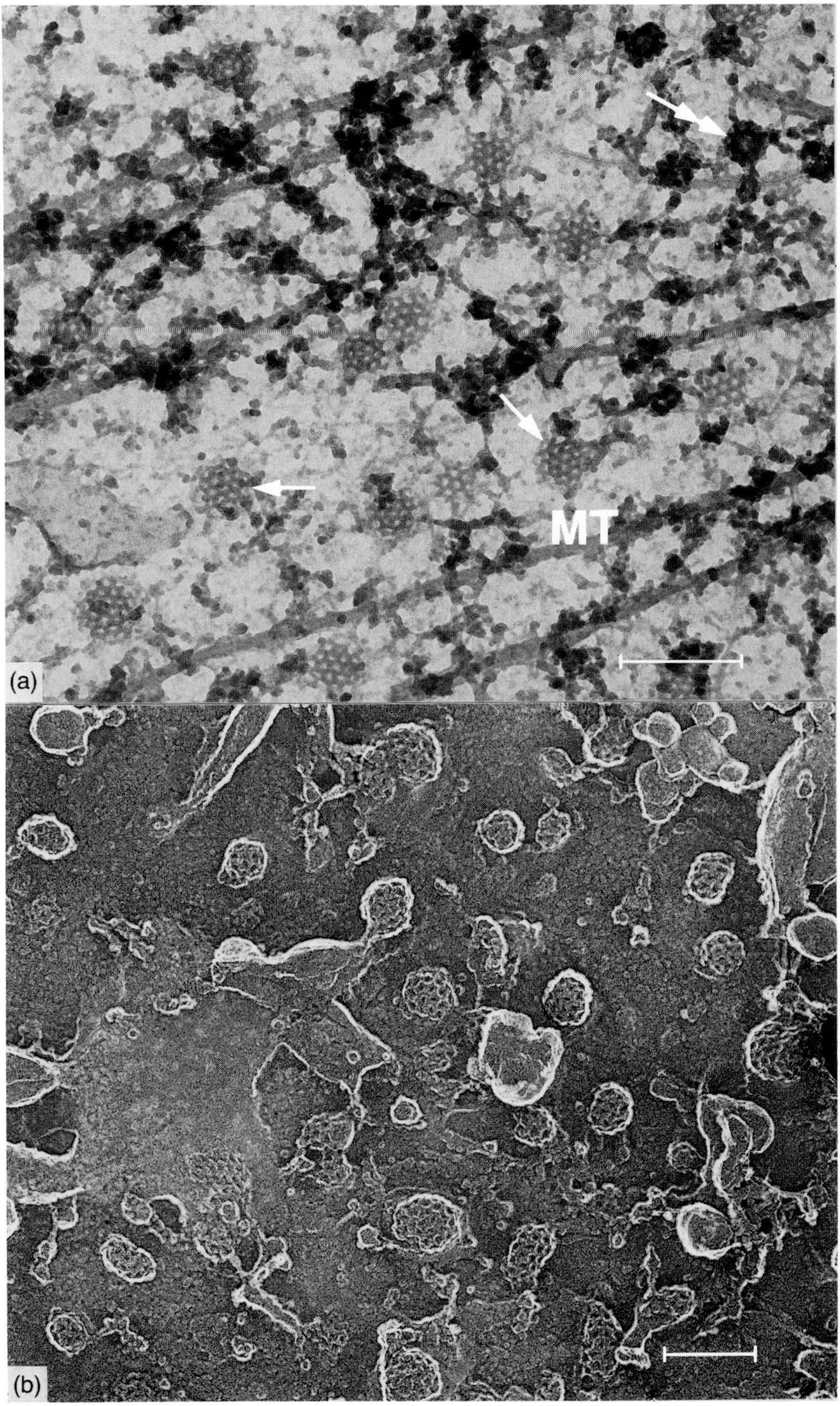
MT
(a)
(b)

Coated vesicle morphology

Coated vesicles are characterised by the structure of the cage or coat which surrounds the membrane-bounded vesicle. The characteristic form of this coat derives from the fact that it is predominantly made up of the protein clathrin. Electron microscopy of the organelle reveals a polygonal basket- or cage-like structure. Each facet of the polygon is either a pentagon or hexagon packed to enclose entirely the vesicle within, giving a dodecahedral appearance (see Figs. 1 and 2). Their structure makes them uniquely identifiable in electron micrographs of appropriate quality and magnification. In higher plants, they are associated with organelles found on the endocytotic and exocytotic pathways, namely the plasma membrane, Golgi apparatus, partially coated reticulum and multivesicular bodies (see Fowke *et al.*, this volume). Coated vesicles associated with the plasma membrane (where they form from coated pits: Tanchak *et al.* 1984; Hillmer *et al.* 1986; Coleman *et al.* 1988) form one distinct population with a diameter of 70–100 nm. A second, smaller population (diameter 40–70 nm) is associated with the Golgi apparatus (Griffing & Fowke 1985; Coleman *et al.* 1987) and partially coated reticulum (Pesacreta & Lucas 1985; Tanchak *et al.* 1988).

Fig. 2. (a) Carbon platinum replica of a coated pit on the inside (PS) face of a carrot suspension-cultured cell. The specimen was prepared by slamming against a helium-cooled copper block, fractured and deep-etched prior to low-angle shadowing and rotary replication. The hexagonal nature of the clathrin cage can be seen clearly, as can the three struts connecting the cage to an adjacent microtubule. The bar represents 50 nm. (From Hawes & Martin 1986, with permission from Academic Press.)

(b) Replica of the inside face of the plasma membrane of a burst carrot protoplast as in Fig. 1(b), showing a fully formed coated vesicle. The bar represents 50 nm. (From Coleman *et. al.* 1987, courtesy of the Company of Biologists Ltd.)

(c) Deep-etch replica of a carrot cell prepared as in (a) showing a clathrin-coated vesicle associated with the Golgi (G) apparatus (arrows). The bar represents 100 nm.

(d) Replica of an ultra-rapidly frozen fractured carrot cell protoplast demonstrating rows of pits in the protoplasmic fracture (PF) face of the plasma membrane. The bar represents 2 μm. (From Coleman *et al.* 1987, courtesy of the Company of Biologists Ltd.)

(e) Replica of a burst protoplast showing a line of coated vesicles apparently associated with a cortical microtubule. The bar represents 100 nm.

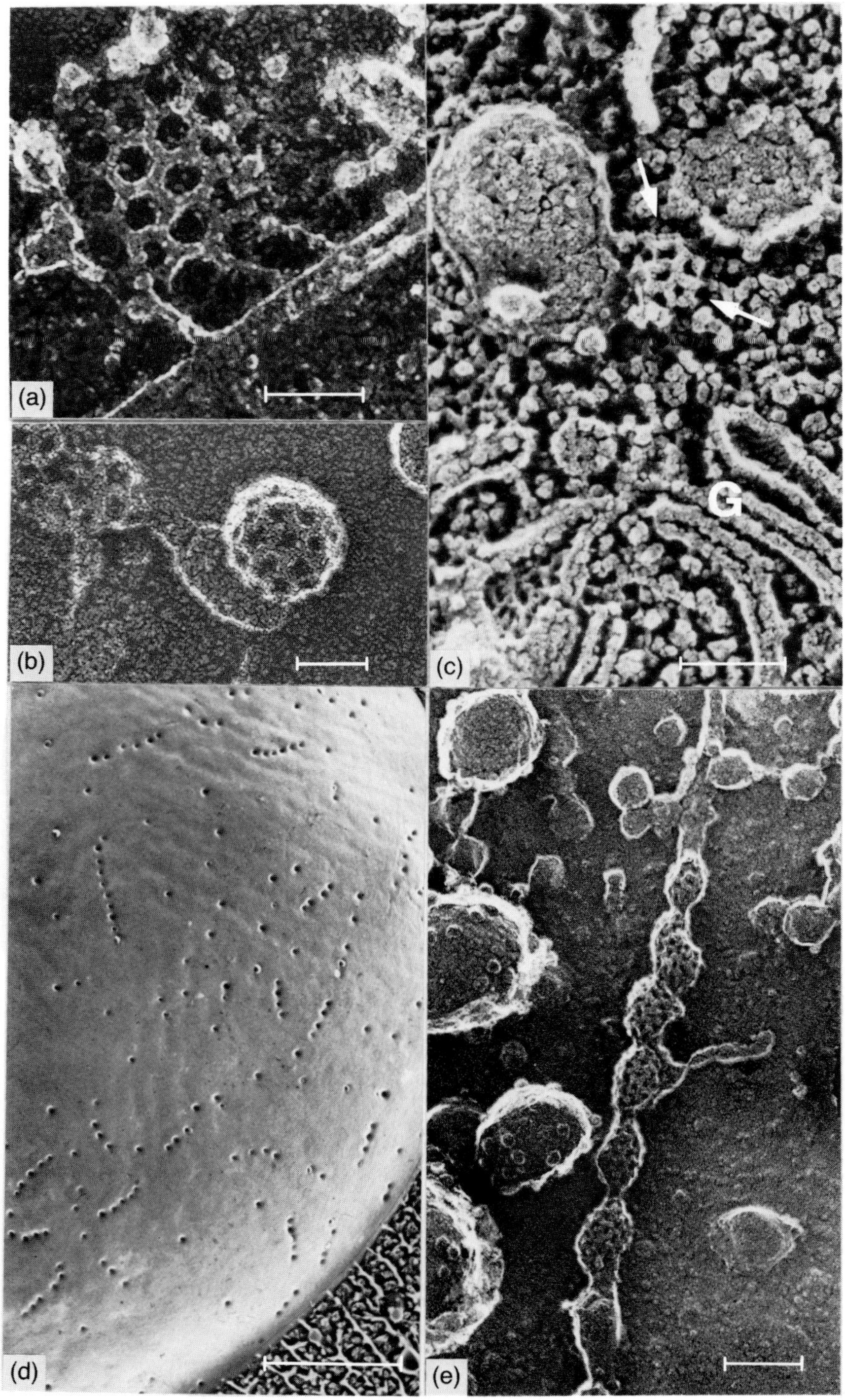
(a)
(b)
(c)
G
(d)
(e)

Apart from the organelles of the endocytotic pathway and the Golgi apparatus there is little evidence of associations between coated vesicles and other cytoplasmic organelles. However, one report (Hawes & Martin 1986) has demonstrated some form of cross-linking between coated pits and cortical microtubules in suspension-cultured carrot cells (Fig. 2(a)). Also, in ultrarapidly frozen and fractured protoplasts from the same cell line, long chains of pits were found in the plasma membrane (Fig. 2(d), and see Coleman *et. al.* 1987) and replicas of the cytoplasmic faces of the plasma membranes occasionally reveal chains of coated vesicles, apparently lined up along microtubules (Fig. 2(e)). However, there is no other evidence to suggest that microtubules are involved in the spatial organisation or transport of plant coated vesicles.

The structure of plant clathrin

Coat purification

Investigations of the molecular components of the vesicle coat and their assembly properties have been attempted largely with isolated coated vesicles. Methods for such isolation and purification of coated vesicles are considered in detail elsewhere (Robinson *et al.*, this volume; Coleman *et al.* 1988) and will not be described here. Sodium dodecyl sulphate/polyacrylamide gel electrophoresis (SDS/PAGE) of such coated vesicle preparations typically reveals many polypeptides which presumably include vesicle contents and contaminants from the preparation (Fig. 3; see also Robinson *et al.*, this volume). In all these preparations a predominant band, the clathrin heavy chain, at M_r 190 000 has been demonstrated (Mersey *et al.* 1985; Depta & Robinson 1986; Depta *et al.* 1987; Coleman *et al.* 1987; Table 1). Having obtained a purified coated vesicle preparation, it is then necessary to isolate coat proteins. In preparations of mammalian clathrin, coat disassembly has been achieved by several methods such as incubation of vesicles with 2 M urea (Ungewickell & Branton 1981) or high ionic strength Tris buffers (e.g. 0.5–1.0 M at pH 6.5–7.0, Keen *et al.* 1979; Kirchausen & Harrison 1981). With plant coated vesicles, similar stripping procedures have been shown to release a variety of polypeptides from the coated vesicle pellet (Weidenhoeft *et al.* 1988; Fig. 4). The number of polypeptides released is dependent on the stripping procedures used and also on the purity of the coated vesicle preparation, but in all cases the M_r 190 000 clathrin band predominates. Further purification of the Tris-released coat has been achieved by precipitation with ammonium sulphate (Coleman *et al.* 1987).

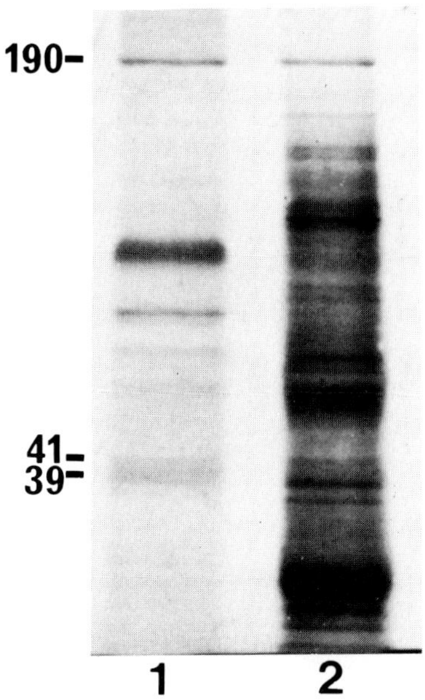

Fig. 3. SDS/PAGE of a carrot coated vesicle preparation. Lane 1 shows proteins eluted from a DEAE-Biogel column after application of coated vesicles isolated on a sucrose density gradient whose protein profile is shown in lane 2. The use of the DEAE column considerably reduces the number of protein bands. The clathrin heavy chain is prominent at M_r 190 000. Bands in the light chain region appear at M_r 39 000 to 41 000.

On the basis of SDS/PAGE the mammalian clathrin heavy chain has a reported relative molecular mass of 180 000 (Keen 1985). This is to be compared with the estimates from equilibrium sedimentation of 170 000 (Pretorius *et al.* 1981) and from estimates derived from the total amino acid sequence deduced from complementary cDNA clones (no. of amino acid residues = 1675, giving an M_r value of 191 569; Kirchhausen *et al.* 1987).

The M_r 190 000 polypeptide in plants shows common antigenicity with the mammalian clathrin heavy chain (see Figs. 5 and 10; Cole *et al.* 1987) from which it differs by about 10 000 in M_r. Immunofluorescence of suspension cultured plant cells with antisera raised against bovine brain clathrin reveals punctate staining in the cortical cytoplasm similar to the classic immunofluorescence of clathrin-coated vesicles in the cytoplasm of mammalian cells such as fibroblasts (Fig. 5 (a) and (b)). Clathrin from other non-mammalian sources (yeast, *Laminaria*, *Chlamydomonas*) has

Table 1. *Comparison of clathrin from different organisms*

	Animal (various authors)	Yeast	*Neurospora* and *Uromyces*	*Laminaria*	*Chlamydomonas*	Higher plants: various authors (see text for refs.)
M_r Heavy chains ($\times 10^{-3}$)	180	185[1], 190[2]	180[3]	170[4]	185[5]	190, 185[6]
M_r light chains ($\times 10^{-3}$)	30–40	36[2]	30–36[3]	36[4]	40[5]	30–38
Triskelion arm (nm)	44–50	49 ± 3.5[1]		33		61 ± 0.99[7]

Notes: [1]Mueller & Branton 1984; [2]Payne & Schekman 1985; [3]That *et al.* 1987; [4]Hang *et al.* 1987; [5]Denning 1989, [6]Wiedenhoeft *et al.* 1988; [7]Coleman *et al.* 1987.

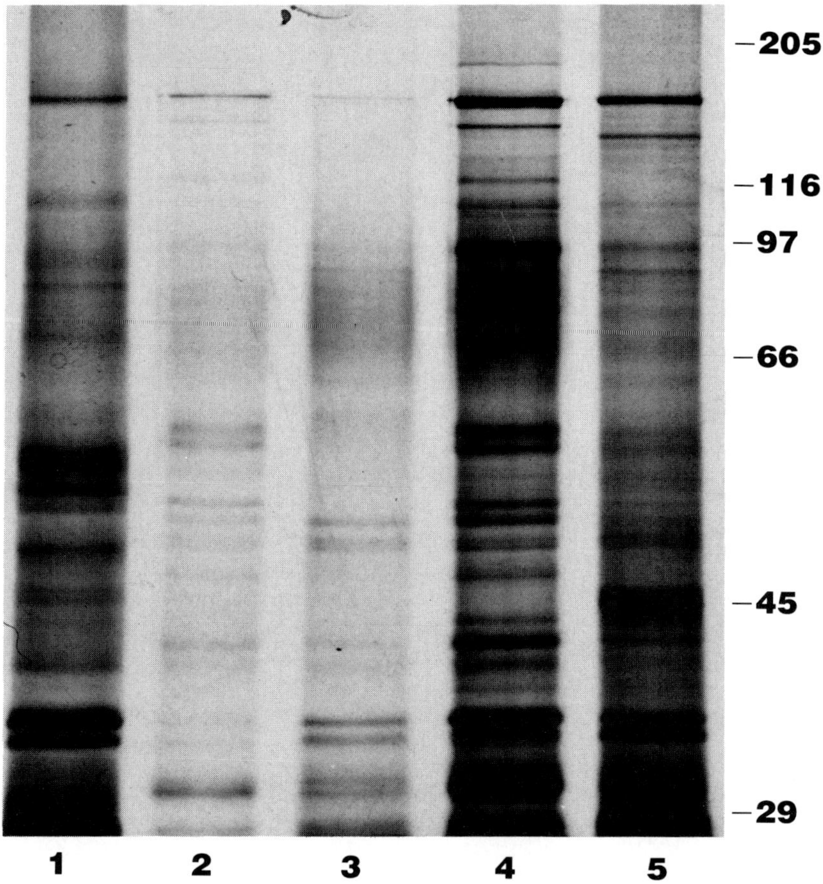

Fig. 4. SDS/PAGE of a carrot coated vesicle preparation subject to various coat releasing agents. Stripped vesicles were pellets at 100 000 G for 1 h and the supernatant run on 7.5% polyacrylamide gells. Lane 1, 0.7 M Tris; lane 2, 2 M urea; lane 3, 1.2 M thiocyanate; lane 4, 0.02 M Taps; lane 5, coated vesicle pellet.

been shown by SDS/PAGE to have M_r values ranging from 170 000 to 190 000 (see Table 1).

Clathrin can also be removed, by using stripping agents, from preparations of plasma membranes prepared by osmotically bursting protoplasts which have been attached to either poly-L-lysine-coated glass slides or coated electron microscope grids. The removal of the clathrin coat from invaginating coated pits can then be monitored by immunofluorescence

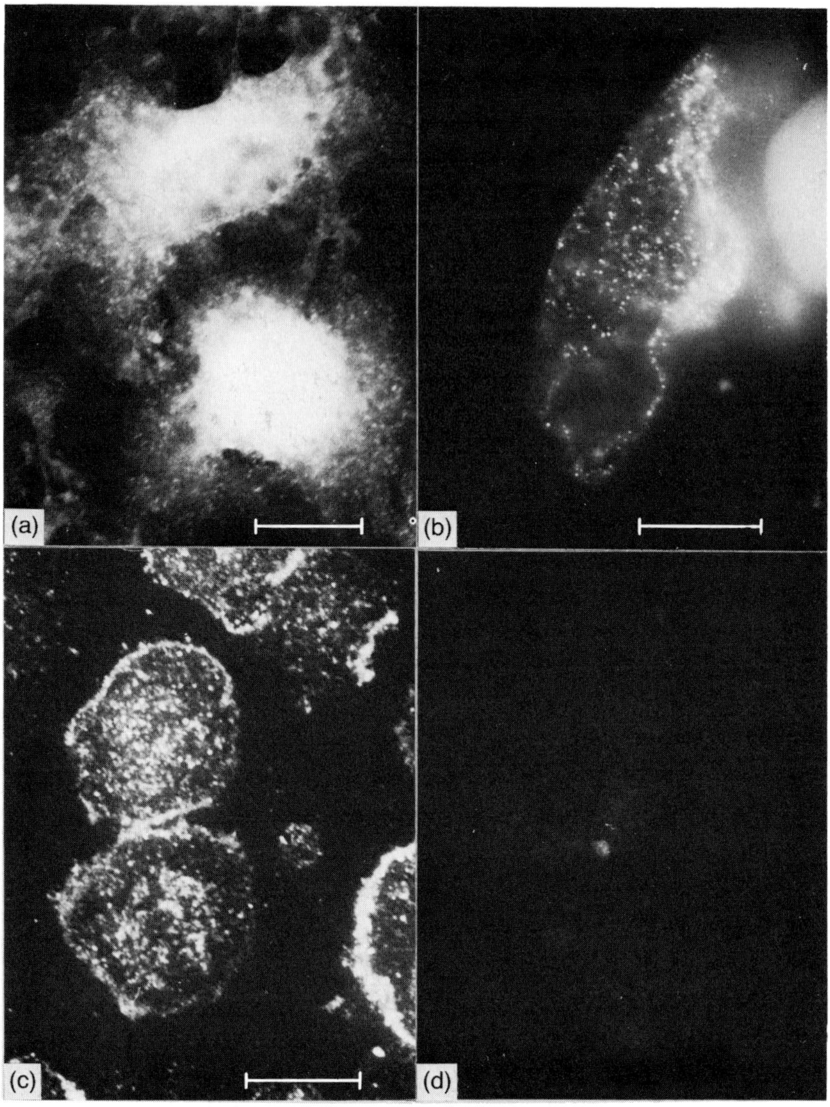

Fig. 5. (a) Immunofluorescence staining of clathrin-coated vesicles in a 3T3 fibroblast. The bar represents 20 μm. (b) Immunofluorescence staining of coated vesicles in a suspension-cultured carrot cell. The bar represents 15 μm. (c) Immunofluorescence staining of coated pits and vesicles on the inside face of a burst protoplast after 12 h of incubation in 10 nM Mes buffer. The bar represents 20 μm. (d) Same burst protoplasts as in (c) but after 12 h of incubation in 0.7 M Tris (pH 6.5). All the clathrin has been removed from the plasma membrane.

(Fig. 5 (c) and (d) or by direct electron microscope observation of critical point-dried preparations (Hawes *et al.* 1989). In these experiments 0.7 M Tris buffer appeared to be the most efficient for dissociation of the clathrin cage.

Electron microscopy of rotary shadowed preparations of plant clathrin prepared by the Tris stripping procedure (Coleman *et al.* 1987) reveal the characteristic three-armed structure, the triskelion (Fig. 6), as reported for mammalian clathrin (Keen 1985). The triskelion has a distinctive morphology (Figs. 6 and 7 (a)); each arm, which radiates from a central point or vertex, is kinked mid way and ends in a globular terminal domain (Coleman *et al.* 1987). In mammalian clathrin, the whole arm is 45–55 nm in length, the kink occurring 16–20 nm from the vertex, separating the proximal from the distal regions of the arm. The amino terminus of the arm lies in the terminal domain, while the carboxyl terminus is directed to the vertex. In the only reported study of plant clathrin triskelions it was reported that the arms are longer than those found typically in mammalian triskelions, with an average length of 61 nm as opposed to 44–50 nm. This correlates with the slightly higher M_r of plant clathrin (Coleman *et al.* 1987).

The three-armed structure of clathrin is the basis for the morphology of the vesicle coat. The observed structure of the coat can be derived by the close packing of the triskelion arms (Fig. 7). The centre of each triskelion lies at the vertex of a polyhedron; each vertex is attached to three surrounding vertices by three struts (approximately 18.5 nm from vertex to vertex). Each strut corresponds to the edge of a polyhedron and consists of two proximal and two distal triskelial arms (Pearse & Crowther 1987). It is interesting to note that the polygon centre-to-centre distance is the same in plant and animal clathrin baskets at around 21 nm; thus it could be assumed that any extra length in the plant triskelial arms must be in the distal and not the proximal portions. The terminal domains are not involved directly in the self-assembly of clathrin to form cages, as their removal by limited proteolysis does not inhibit cage formation (Ungewickell *et al.* 1982) but may be involved in attachment of the clathrin to the plasma membrane (Heuser & Keen 1988) as they have been shown to bend inwards away from the cage and towards the membrane of the vesicle.

The structure formed by the interaction of clathrin molecules is somewhat flexible in nature: the triskelions can interact to form either pentagons or hexagons. The assembly of the clathrin cage obeys Euler's theorem, which predicts that in order to achieve the necessary curvature to form a closed three-dimensional cage then a minimum of 12 pentagons are required, with a variable number of hexagons inserted between them.

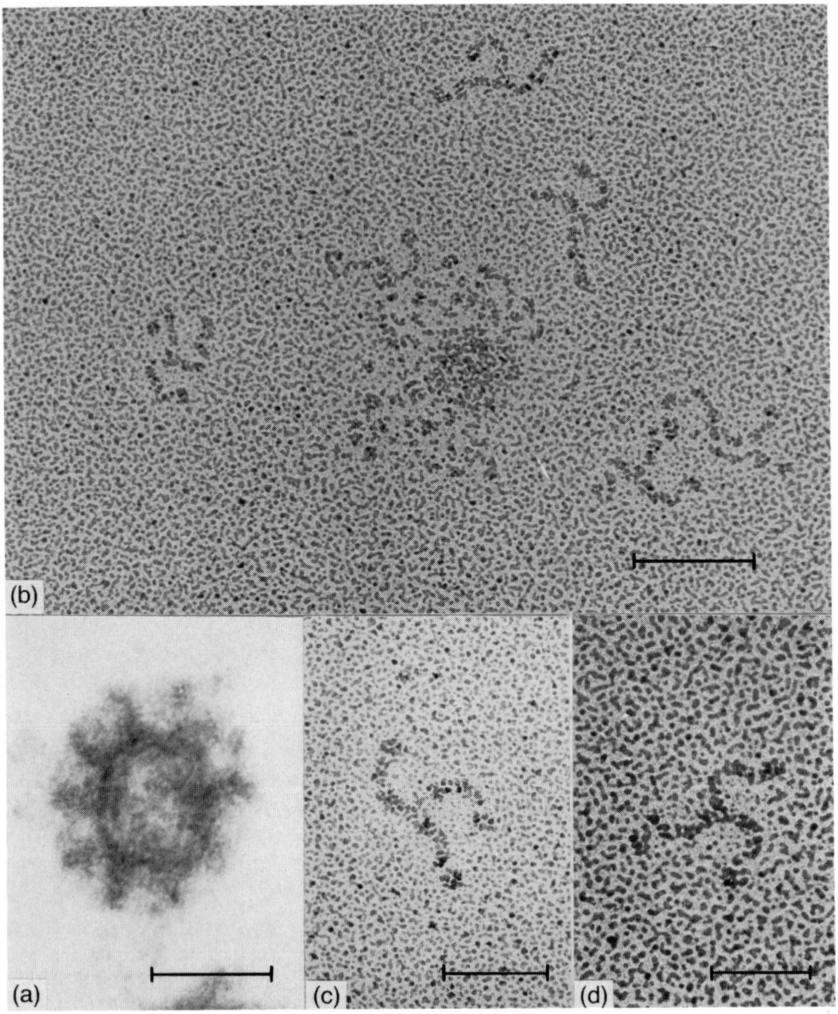

Fig. 6. (a) High magnification of an isolated coated vesicle. The bar represents 50 nm. (b) Rotary-shadowed clathrin triskelions released from carrot coated vesicles with Tris and precipitated with ammonium sulphate prior to spraying on to mica sheets, followed by rotary shadowing. The bar represents 100 nm.
(c) and (d) Clathrin triskelions. The bars represent 50 nm. (From Coleman *et al.* 1987, with permission from the Company of Biologists Ltd.)

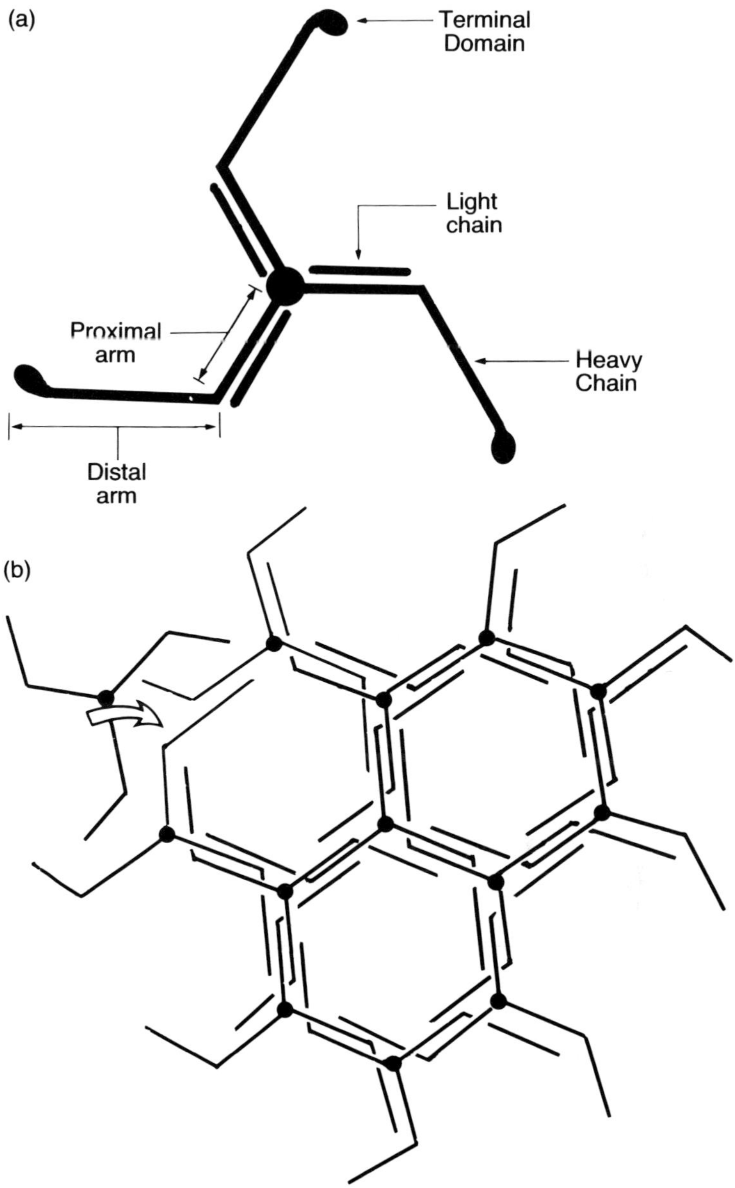

Fig. 7. (a) The structure of the clathrin triskelion. It is thought that the terminal domain of the heavy chain bends in towards the vesicle membrane and binds to receptors via adaptor molecules. (b) Packing of the triskelions to form the clathrin cage. (From Coleman *et al.* 1988, courtesy of Blackwells Scientific Publications.)

This then forms the classic pentagonal dodecahedron, which is a structure very common in nature, found for example in various virus capsids and insect bronchosomes (Galloway 1988).

Light chains

In addition to the heavy-chain polypeptides, soluble clathrin also comprises lower relative molecular mass light chains. In vertebrates these light-chain polypeptides are heterogeneous, with M_r values varying from 30 000 to 40 000. There are two major classes of light chain, LC_a and LC_b, coded for by separate genes in single copies, and showing 60% sequence identity. Two sub-types of LC_a and one sub-type of LC_b have been identified in vertebrate brain tissue and differ from the LC_a and LC_b sub-types of non-brain tissues by specific insertion sequences (Brodsky & Parham 1983; Kirchhausen *et al.* 1987). Conserved regions of the light chains appear to be located in the heavy-chain binding domain (Brodsky *et al.* 1987; Blank & Brodsky 1987).

Although SDS/PAGE gels of higher plant coated vesicles show a number of polypeptides in the M_r 30 000 to 40 000 regions, positive identification of light chains has proved to be more difficult (Robinson & Depta 1988). No polypeptides in the M_r 30 000 to 40 000 range cross-reacted with antibodies to mammalian light chains (Cole *et al.* 1987; Balusek *et al.* 1988). The doublet in the M_r 57 000 range which cross-reacted with antibodies raised against bovine light chains were tentatively identified as possible candidates for plant light chains (Cole *et al.* 1987). However, it now seems unlikely that these are light chains; they probably represent proteolytic degradation products of the clathrin heavy chain.

Two features of mammalian light chains, their heat stability and ability to bind calcium (Brodsky 1988), have been used to identify tentatively light chains of M_r 30 000 and 38 000 in bean coated vesicles (Balusek *et al.* 1988). Denning (1989) has demonstrated also an M_r 40 000 protein band in SDS/PAGE gels of *Chlamydomonas* coated vesicles which was heat stable (Fig. 8, lane 1), sensitive to low concentrations of elastase (Fig. 8, lanes 3–5) (both features of mammalian light chains; Ungewickell *et al.* 1982) and had a 1:1 stoichiometry with the heavy chain. The larger M_r value of the light chain in this system underlines the need to identify light chains by criteria other than this property alone.

A single polypeptide with M_r 36 000 has been identified in purified clathrin from yeast (Mueller & Branton 1984; Payne & Schekman 1985; see also Table 1) and the gene has been located which shows strong homology with the conserved regions of vertebrate light chains (Silvera *et al.* 1988). The role of light chains is unclear; they are not essential for coat

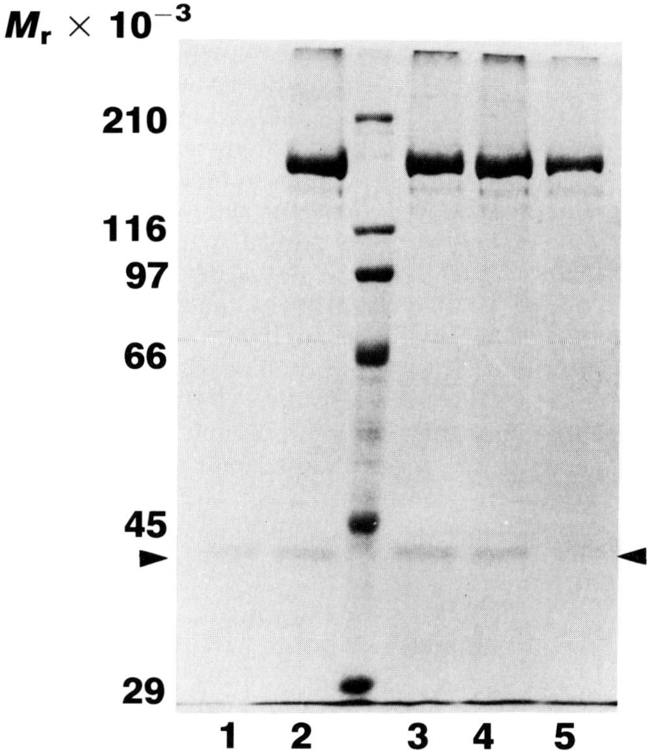

Fig. 8. SDS/PAGE of a *Chlamydomonas* coated vesicle pellet, demonstrating the heat resistance and elastase sensitivity of putative M_r 40 000 light chains. Lane 1, heat-treated vesicle pellet; lane 2, untreated coated vesicles; lanes 3–5, 0.05, 0.1, 0.5 μg elastase ml^{-1}. (Figure kindly supplied by Dr Gerene Denning.)

assembly but are required for the enzymic uncoating by ATPase and they appear to increase the phosphorylation of the 50 kDa coat assembly protein (Brodsky 1988).

Assembly and disassembly

Isolated plant and animal triskelions, at critical concentrations, reassemble spontaneously into empty polygonal cages (Keen *et al.* 1979; Coleman *et al.* 1987; Wiedenhoft *et al.* 1988). Assembly *in vitro* does not require energy but can be regulated by changes in ionic strength including Ca^{2+} concentration and pH (Brodsky 1988). Such self-assembly of mammalian clathrin coats under physiological conditions *in vitro* requires the presence of a group of polypeptides, present in coats *in vivo*, termed 'assembly factors', 'assembly polypeptides' (Ahle & Ungewickell 1986; Ahle *et al.* 1988; Keen *et al.* 1979; Prasad & Lippold 1988; Zaremba & Keen 1983) or adaptors (Pearse 1988; Robinson 1987, 1989).

There are at least two sorts of adaptor proteins, HA1 and HA2. Both are multi sub-unit proteins each containing heterodimers from a family of M_r 110 000 to 116 000 polypeptides termed adaptins (Keen *et al.* 1979; Pearse & Robinson 1984; Morris *et al.* 1989) plus two smaller polypeptides of M_r 47 000 and 20 000 in HA1 and M_r 50 000 and 16 000 in HA2 (Morris *et al.* 1989). The function of the adaptor complex is to link the clathrin, probably via the terminal domains, to receptor proteins in the vesicle membrane (Pearse & Crowther 1987; Heuser & Keen 1988; Fig. 9). Sequence information in the cytoplasmic tails of receptors specifies the interaction of the appropriate adaptor (Glickman *et al.* 1989). One adaptor, HA1, is Golgi specific, whilst HA2 is specific for plasma-

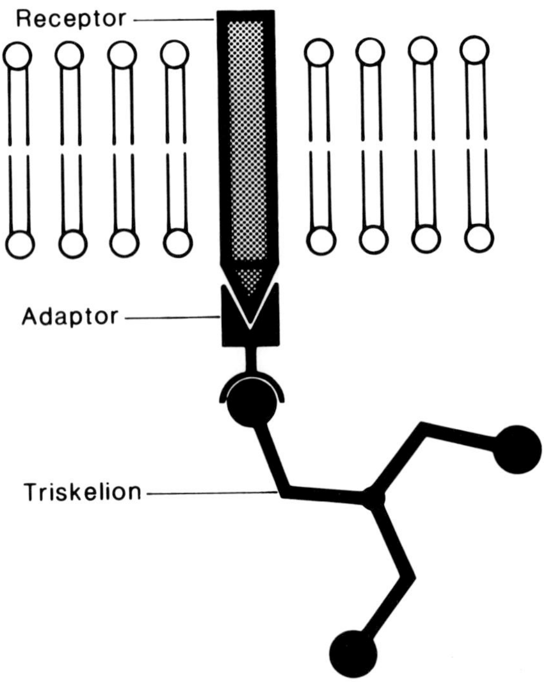

Fig. 9. Diagram illustrating the role of adaptor proteins in linking the terminal domain of the clathrin triskelion with a transmembrane receptor.

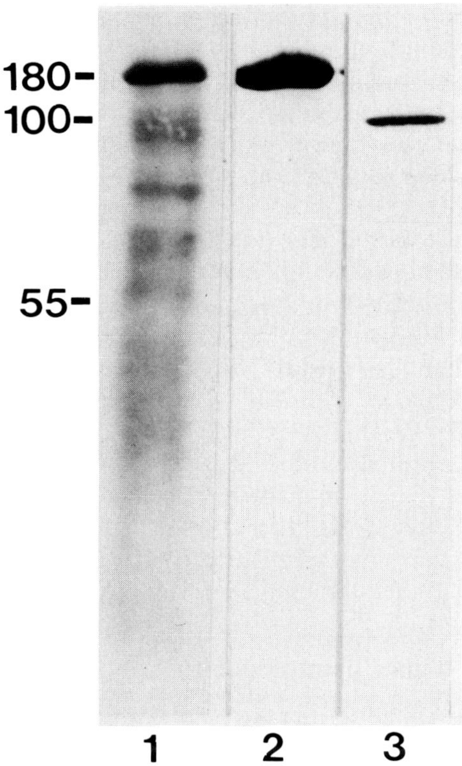

Fig. 10. Putative soybean assembly polypeptide. A soybean coated vesicle preparation (lane 1) was Western blotted against an anti-bovine brain clathrin heavy-chain polyclonal (lane 2) and against an anti-bovine brain 110 000 M_r clathrin-associated protein. The numbers are $M_r \times 10^{-3}$. (Figure kindly supplied by Dr Steve Moseley.)

membrane-located receptors (Robinson 1987; Ahle *et al.* 1988; Morris *et. al.* 1989), thus giving the cell the ability to sort clathrin between the plasma membrane and Golgi apparatus.

SDS/PAGE gels of higher plant coated vesicles sometimes show polypeptides of an M_r similar to those of the subunits of mammalian adaptor proteins (see Robinson & Depta 1988). Tentative identification of an assembly polypeptide amongst these in plant extracts has been made (S.T. Mosley, personal communication). A polyclonal antibody raised against a bovine brain assembly factor (Mosley & Branton 1988) is shown to recognise an M_r 110 000 band in an extract of soybean coated vesicles (see Fig. 10). Firm identification awaits purification of this polypeptide and demonstration that it promotes the reassembly of clathrin baskets from purified clathrin heavy chains *in vitro*. It is possible that protocols developed to date for the isolation and purification of plant clathrin have precluded the isolation of these clathrin-associated proteins. It should be pointed out that we still await any evidence of the interaction of a receptor, be it intracellular or plasma membrane bound, with higher plant clathrin or plant coated vesicles.

Coated vesicles cannot fuse with other cytoplasmic membranes without first losing their coat (Altshiel & Branton 1983) and uncoating is therefore essential to their function. Clathrin dissociation does not occur spontaneously under physiological conditions as the process is thermodynamically unfavourable. However, an ATPase which is responsible for mediating uncoating *in vitro* has been demonstrated to be present, in mammalian cell cytosol (Patzer *et al.* 1982; Schlossman *et. al.* 1984). This enzyme was identified to be one of the hsp70 family of M_r 70 000 heat shock proteins (Rothman & Schmid 1986). It seems likely that a similar uncoating ATPase exists in plant cells; certainly some such system must exist, as the uncoating of plant coated vesicles is just as energetically unfavourable as that of mammalian vesicles.

The functions of higher plant clathrin

In the 15 years since Pearse (1975) first isolated porcine brain coated vesicles a vast number of investigations has yielded important data on the function of mammalian clathrin and its associated proteins. Despite a rather slower start to research into plant coated vesicles, the significant progress described in this chapter and elsewhere (Fowke *et al.* and Robinson *et. al.*, this volume) indicate that we are now in a strong position to characterise and describe the function of plant clathrin and its associated proteins.

It is now clear that clathrin is a vitally important molecule in all eukaryotic cells. Its primary function (at least in animal cells) is to concentrate and sort membrane proteins prior to their transport from one cellular compartment to another within vesicles. Given the conservation of clathrin genes between animals and yeast and the apparent homology with plant clathrin (as predicted from immunological and structural data) it is highly probable that there is an equivalent homology of function across the kingdoms.

The action of clathrin is dependent upon its adaptor complexes differentiating between specific sets of receptors at the plasma membrane and *trans*-Golgi. So far identification of such proteins in plants has been hindered by problems of the high proteolytic activity which is encountered in plant extracts. Such problems and the development of reliable purification protocols are discussed by Robinson *et al.* (this volume). However, it is tempting to speculate that in plant cells the mechanism of receptor-mediated endocytosis via clathrin-coated vesicles exists. Verification of this awaits the detection of receptor internalisation into plant coated vesicles. Future research therefore lies in two major areas: firstly in the study of plant clathrin itself and secondly in the isolation and characterisation of membrane-bound receptors that are candidates for internalisation into plant coated pits. Perhaps the key to both of these goals will be the identification of the plant equivalent of the adaptor polypeptides linking clathrin to transmembrane receptors.

It can be concluded that it is still early days in the study of plant clathrin and coated vesicles, and much work still awaits us before the precise roles of these small organelles are revealed. The application of modern immunological techniques together with improvements in coated vesicle purification will permit the identification of clathrin-associated proteins and the elucidation of their function. Studies using mutation and genetic manipulation will give data on gene products essential to coated vesicle function and provide an alternative approach to the analysis of their function.

References

Ahle, S., Mann, A., Eichelsberger, U. & Ungewickell, E. (1988). Structural relationships between clathrin assembly proteins from the Golgi and plasma membrane. *EMBO Journal*, **7**, 919–29.

Ahle, S. & Ungewickell, E. (1986). Purification and properties of a new clathrin assembly protein. *EMBO Journal*, **5**, 3143–9.

Altshiel, L. & Branton, D. (1983). Fusion of coated vesicles with lysosomes: measurement with a fluorescence assay. *Cell*, **32**, 921–9.

Balusek, K., Depta, H. & Robinson, D.G. (1988). Two polypeptides (30 and 38 kDa) in plant coated vesicles with clathrin light chain properties. *Protoplasma*, **146**, 174–6.

Barrosa, J. & Pais, M.S.S. (1987). Coated vesicles in the cytoplasm of the host cells in *Ophrys lutae* Cav. mycorrhizas (Orchidacea). *New Phytologist*, **105**, 67–70.

Blank, G.S. & Brodsky, F.M. (1987). Clathrin assembly involves a light chain-binding region. *Journal of Cell Biology*, **105**, 2011–19.

Brodsky, F.M. (1988). Living with clathrin: its role in intracellular membrane traffic. *Science*, **242**, 1396–1402.

Brodsky, F.M., Galloway, C.J., Blank, G.S., Jackson, A.P., Seow, H.-F., Drickamer, K. & Parham, P. (1987). Localization of clathrin light chain sequences mediating heavy chain binding and coated vesicle diversity. *Nature (London)*, **326**, 203–5.

Brodsky, F.M. & Parham, P.J. (1983). Polymorphism in clathrin light chains from different tissues. *Journal of Molecular Biology*, **167**, 197–204.

Cole, L., Coleman, J.O.D.C., Evans, D.E., Hawes, C.R. & Horsley, D. (1987). Antibodies to brain clathrin recognise plant coated vesicles. *Plant Cell Reports*, **6**, 227–30.

Coleman, J., Evans, D. & Hawes, C. (1988). Plant coated vesicles. *Plant, Cell and Environment*, **11**, 669–84.

Coleman, J., Evans, D.E., Hawes, C.R., Horsley, D. & Cole, L. (1987). Structure and molecular organisation of higher plant coated vesicles. *Journal of Cell Science*, **88**, 35–45.

Denning, G.M.C. (1989). Studies of the structure and function of clathrin-coated vesicles in *Chlamydomonas*, a unicellular green alga. Ph.D. thesis, University of Iowa.

Depta, H., Freundt, H., Hartmann, D. & Robinson, D.G. (1987). Preparation of a homogeneous coated vesicle fraction from bean leaves. *Protoplasma*, **136**, 154–60.

Depta, H. & Robinson, D.G. (1986). The isolation and enrichment of coated vesicles from suspension-cultured carrot cells. *Protoplasma*, **130**, 162–70.

Emons, A.M.C. & Traas, J.A. (1986). Coated pits and coated vesicles on the plasma membrane of plant cells. *European Journal of Cell Biology*, **41**, 57–64.

Esau, K. & Gill, R.H. (1970). Observations on spiny vesicles and P-protein in *Nicotiana tabacum*. *Protoplasma*, **69**, 373–88.

Galloway, J. (1988). Nature's second-favourite structure. *New Scientist*, **1606**, 36–9.

Glickman, J.N., Konibear, E. & Pearse, B.M.F. (1989). Specificity of binding of clathrin adaptors to signals on the mannose-6-phosphate insulin-like growth factor-II receptor. *EMBO Journal*, **8**, 1041–7.

Griffing, L.R. & Fowke, L.C. (1985). Cytochemical localization of peroxidase in soybean suspension culture cells and protoplasts: intracellular vacuole differentiation and presence of peroxidase in coated vesicles and multivesicular bodies. *Protoplasma*, **128**, 22–30.

Hang, X.H., Bomsel, M., de Paillerets, C., Weintraub, H. & Alfsen, A. (1987). Homologie entre des déterminants antigéniques de clathrine de boeuf et de clathrine d'une algu brune *Laminaria digitata*. *Compte Rendu de l'Académie des Sciences Paris*, **304**, 465–8.

Harley, S.M. & Beevers, L. (1989). Coated vesicles are involved in the transport of storage proteins during seed development in *Pisum sativum* L. *Plant Physiology*, **91**, 674–8.

Hawes, C.R. (1985). Conventional and high voltage electron microscopy of the cytoskeleton and cytoplasmic matrix of carrot (*Daucus carota* L.) cells grown in suspension culture. *European Journal of Cell Biology*, **38**, 201–10.

Hawes, C., Coleman, J., Evans, D. & Cole, L. (1989). Recent advances in the study of plant coated vesicles. *Cell Biology International Reports*, **13**, 119–28.

Hawes, C. & Martin, B. (1986). Deep etching of plant cells: cytoskeleton and coated pits. *Cell Biology International Reports*, **10**, 985–92.

Heuser, J.E. & Keen, J. (1988). Deep-etch visualisation of proteins involved in clathrin assembly. *Journal of Cell Biology*, **107**, 877–86.

Hillmer, S., Depta, H. & Robinson, D.G. (1986). Confirmation of endocytosis in higher plant protoplasts using lectin–gold conjugates. *European Journal of Cell Biology*, **41**, 142–9.

Keen, J.H. (1985). The structure of clathrin coated membranes: assembly and disassembly. In *Endocytosis*, ed. I. Pastan & M. Willingham, pp. 85–126. New York: Plenum Press.

Keen, J.H., Willingham, M.C. & Pastan, I.H. (1979). Clathrin coated vesicles: isolation, dissociation and factor-dependent reassociation of clathrin baskets. *Cell*, **16**, 303–12.

Kirchhausen, T. & Harrison, J.R. (1981). Protein organisation in clathrin trimers. *Cell*, **23**, 755–61.

Kirchhausen, T., Harrison, S.C., Chow, P.E., Mattaliano, R.J., Ramachandran, K.L., Smart, J. & Brosius, J. (1987). Clathrin heavy chain: molecular cloning and complete primary structure. *Proceedings of the National Academy of Sciences, USA*, **84**, 8805–9.

Kirchhausen, T., Scarmato, P., Harrison, S.C., Monroe, J.J., Chow, E.P., Mattaliano, R.J., Ramachandran, K.L., Smart, J.E., Ahn, A.H. & Brosius, J. (1987). Clathrin light chains LCA and LCB are similar, polymorphic and share repeated heptad motifs. *Science*, **236**, 320–4.

Mersey, B.G., Griffing, L.R., Rennie, P.J. & Fowke, L.C. (1985). The isolation of coated vesicles from protoplasts of soybean. *Planta*, **163**, 317–27.

Morris, S.A., Ahle, S. & Ungewickell, E. (1989). Clathrin-coated vesicles. *Current Opinion in Cell Biology*, **1**, 684–90.

Mosley, S.T. & Branton, D. (1988). An antibody against 100- to 116-kDa polypeptides in coated vesicles inhibits triskelion binding. *Experimental Cell Research*, **174**, 511–20.

Mueller, S. & Branton, D. (1984). Identification of coated vesicles in *Saccharomyces cerevisiae*. *Journal of Cell Biology*, **98**, 341–6.

Nakamura, S. & Miki-Hirosige, H. (1982). Coated vesicles and cell plate formation in the microspore mother cell. *Journal of Ultrastructure Research*, **80**, 302–11.

Newcomb, E.H. (1980). Coated vesicles: their occurrence in different plant cell types. In *Coated Vesicles*, ed. C.D. Ockleford & A. Whyte, pp. 55–68. Cambridge: Cambridge University Press.

Ockleford, C.D. & Whyte, A. (1980). *Coated Vesicles*. Cambridge, Cambridge University Press.

Patzer, E.J., Schlossman, D.M. & Rothman, J.E. (1982). Release of clathrin from coated vesicles dependent upon a nucleoside triphosphate and a cytosol fraction. *Journal of Cell Biology*, **93**, 230–6.

Payne, G.S. & Schekman, R. (1985). A test of clathrin function in protein secretion and cell growth. *Science*, **230**, 1009–12.

Pearse, B.M.F. (1975). Coated vesicles from pig brain: purification and biochemical characterisation. *Journal of Molecular Biology*, **97**, 93–8.

Pearse, B.M.F. (1988). Receptors compete for adaptors found in plasma membrane coated pits. *EMBO Journal*, **7**, 3331–6.

Pearse, B.M.F. & Crowther, R.A. (1987). Structure and assembly of coated vesicles. *Annual Review of Biophysics and Biophysical Chemistry*, **16**, 49–68.

Pearse, B.M.F. & Robinson, M.S. (1984). Purification and properties of 100 kd proteins from coated vesicles and their reconstituton with clathrin. *EMBO Journal*, **3**, 1951–7.

Pesacreta, Th. C. & Lucas, W.J. (1985). Presence of a partially coated-reticulum in angiosperms. *Protoplasma*, **125**, 173–84.

Prasad, K. & Lippold, R.E. (1988). Molecular characterization of the AP180 coated vesicle assembly protein. *Biochemistry*, **27**, 6098–7014.

Pretorius, H.T., Nandi, B.K., Lippoldt, R.E., Johnson, M.L., Keen, J.H., Pastan, I. & Edelhoch, J. (1981). Molecular characterization of human clathrin. *Biochemistry*, **20**, 2777–82.

Robertson, J.G. & Lyttleton, P. (1982). Coated and smooth vesicles in the biogenesis of cell walls, plasma membranes, infection threads and peribacteriod membranes in root hairs and nodules of white clover. *Journal of Cell Science*, **58**, 63–78.

Robinson, D.G. & Depta, H. (1988). Coated vesicles. *Annual Review of Plant Physiology and Plant Molecular Biology*, **39**, 53–99.

Robinson, M.S. (1987). 100-kD coated vesicle proteins: molecular heterogeneity and intracellular distribution studies with monoclonal antibodies. *Journal of Cell Biology*, **104**, 887–95.

Robinson, M.S. (1989). Cloning of cDNAs encoding two related 100-kD coated vesicle proteins (α-adaptins). *Journal of Cell Biology*, **108**, 833–42.

Rothman, J.E. & Schmid, S.L. (1986). Enzymatic recycling of clathrin from coated vesicles. *Cell*, **46**, 5–9.

Ryser, U. (1979). Cotton fibre differentiation: occurrence and distribution of coated and smooth vesicles during primary and secondary wall formation. *Protoplasma*, **98**, 223–39.

Schlossman, D.M., Schmid, S.L., Braell, W.A. & Rothman, J.E. (1984). An enzyme that removes clathrin coats: purification of an uncoating ATPase. *Journal of Cell Biology*, **99**, 723–33.

Sheldon, J.M. & Dickinson, H.G. (1983). Determination of patterning in the pollen wall of *Lilium henryi*. *Journal of Cell Science*, **63**, 191–208.

Silvera, L., Wang, D., Masiarz, F. & Schekman, R. (1988). Identification of the clathrin light chain gene from *Saccharomyces cerevisiae*. *Journal of Cell Biology*, **107**, 737.

Tanchak, M.A., Griffing, L., Mersey, B. & Fowke, L. (1984). Endocytosis of cationised ferritin by coated vesicles of soybean protoplasts. *Planta*, **162**, 481–6.

Tanchak, M.A., Rennie, P.J. & Fowke, L.C. (1988). Ultrastructure of the partially coated reticulum and dictyosomes during endocytosis by soybean protoplasts. *Planta*, **175**, 433–41.

That, T.C., Hoang-Van, K., Turian, G. & Hoch, H.C. (1987). Isolation and characterisation of coated vesicles from filamentous fungi. *European Journal of Cell Biology*, **43**, 189–94.

Ungewickell, E. & Branton, D. (1981). Assembly units of clathrin coats. *Nature (London)*, **239**, 420–2.

Ungewickell, E., Unanue, E.R. & Branton, D. (1982). Functional and structural studies on clathrin triskelions and baskets. *Cold Spring Harbor Symposium on Quantitative Biology*, **46**, 723–31.

Wiedenhoeft, R.E., Schmidt, G.W. & Palevitz, B.A. (1988). Dissociation and reassembly of soybean clathrin. *Plant Physiology*, **86**, 412–16.

Zaremba, S. & Keen, J.J. (1983). Assembly polypeptides from coated vesicles mediate reassembly of unique clathrin coats. *Journal of Cell Biology*, **97**, 1339–47.

D. G. ROBINSON, K. BALUSEK, H. DEPTA,
B. HOH AND S. E. H. HOLSTEIN

Isolation and characterisation of plant coated vesicles

Isolation of plant coated vesicles

Coated vesicles (CVs) are the 'world's smallest organelles', if the definition of an organelle as a membrane-bounded compartment can be stretched this far. Their size, less than 100 nm in diameter, means that they can be treated essentially as viruses in terms of their isolation. Because antibodies against plant clathrin are not yet readily available for immunoaffinity separations, the isolation of CVs from plant sources has so far been carried out exclusively by centrifugal means. This means that we are dealing with a subcellular particle belonging to the postmicrosomal supernatant.

Methods for isolation and purity of fractions obtained

There are a number of centrifugal procedures by which CVs can be isolated from homogenates of animal cells, but only a few of these have been used with success on plants. An important difference between plants and animals, in terms of CV isolation, is the relatively high amount of ribosomes in the plant postmicrosomal pellet. This makes some isolation protocols, e.g. those of Keen *et al.* (1979) or Campbell *et al.* (1983), completely unsuitable for the plant cell biologist. In fact 'CV-fractions' prepared by the latter two isolation protocols from zucchini hypocotyl tissue are essentially ribosome preparations (Fig. 1 (a) and (b); see also Mersey *et al.* 1982)!

In 1986 Depta & Robinson introduced a procedure for CV isolation from higher plants which, by treating the postmicrosomal pellet with RNase, overcomes the problem of ribosomal contamination. This method works equally well on both etiolated and green (e.g. Depta *et al.* 1987*a*) tissue, and has been adopted by other groups (Coleman *et al.* 1987; Harley & Beevers 1989*a*). It incorporates features of the isolation protocol of Pearse (1982) for CVs from human placenta and involves, in

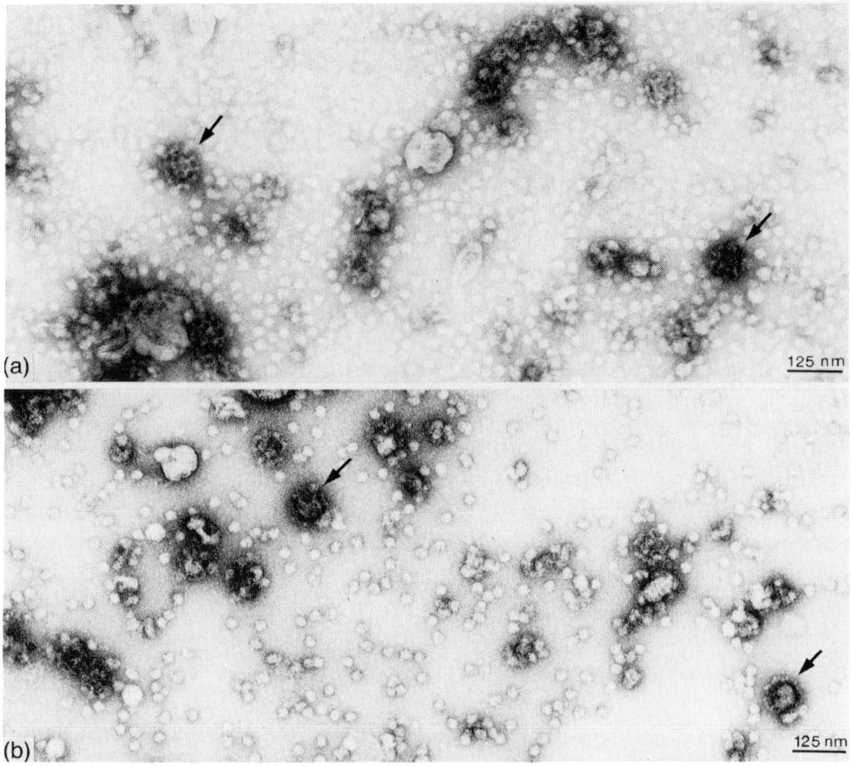

Fig. 1. Micrographs of negatively stained 'coated vesicle' (arrowed) fractions from zucchini hypocotyls prepared according to the procedure of (a) Keen *et al.* (1979), and (b) Campbell *et al.* (1983). These methods were introduced for CV isolation from bovine brain and rat liver, respectively. In plants, these fractions are grossly contaminated with ribosomes.

addition to the RNase treatment, a short high-speed sedimentation on to a sucrose cushion, followed by an isopycnic centrifugation on a linear $^{2}H_2O$/Ficoll gradient.

The protein distribution on such a gradient is presented in Fig. 2 and, for isolations from fresh material, reveals two peaks. The larger one, termed region I by Depta & Robinson (1986), banding at around 9% (w/w) sucrose equivalents, is composed mainly of smooth membranes; the smaller one, region II, contains CVs. Depending on exactly which fractions are pooled to create region II, contamination with smooth membranes can vary from 15% to 30% as judged by particle counting in negatively stained preparations. These contaminants can be almost

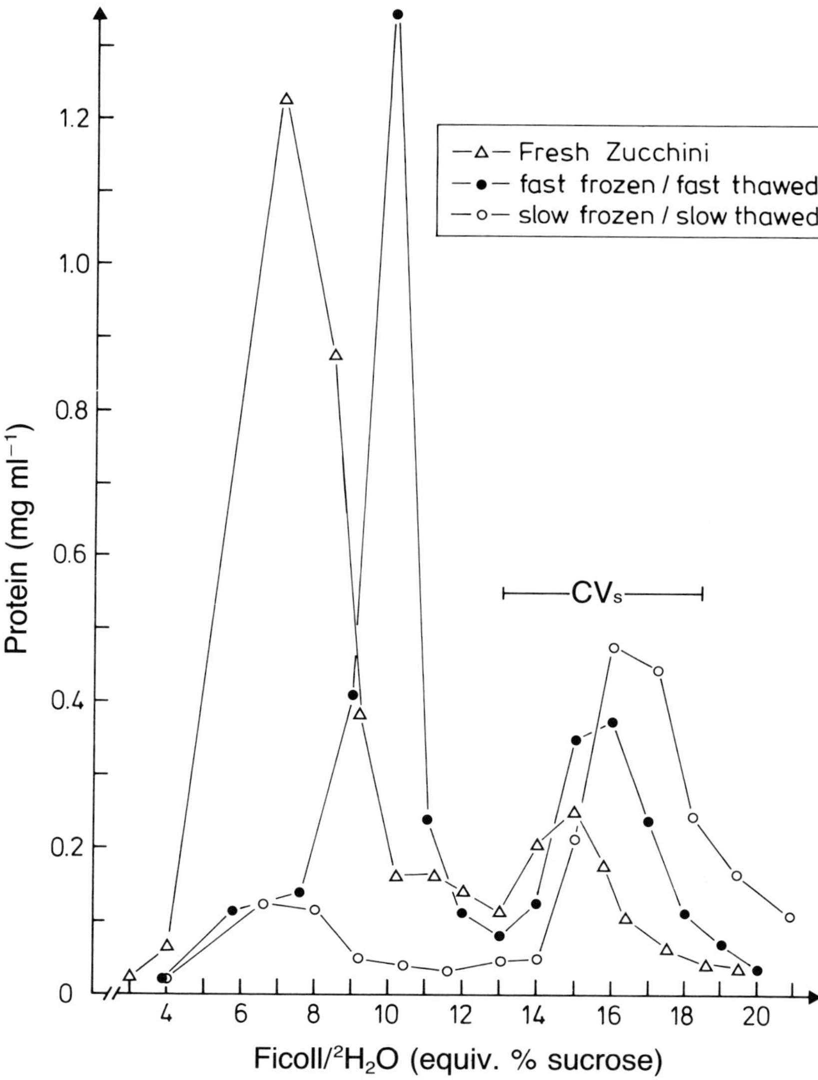

Fig. 2. Protein profiles of RNase-treated, postmicrosomal fractions from zucchini hypocotyls centrifuged isopycnically in Ficoll/^2H_2O gradients according to the method of Depta & Robinson (1986). The CV-containing fractions (region II) are indicated with a bracket. Note that freezing leads to a slight increase in CV density and that region I (6% to 10% equiv. sucrose) almost disappears in the case of slow-frozen (placed in a freezer at −80 °C), slow-thawed (overnight in a refrigerator at +4 °C) material. Amount of starting material in each case was 1 kg.

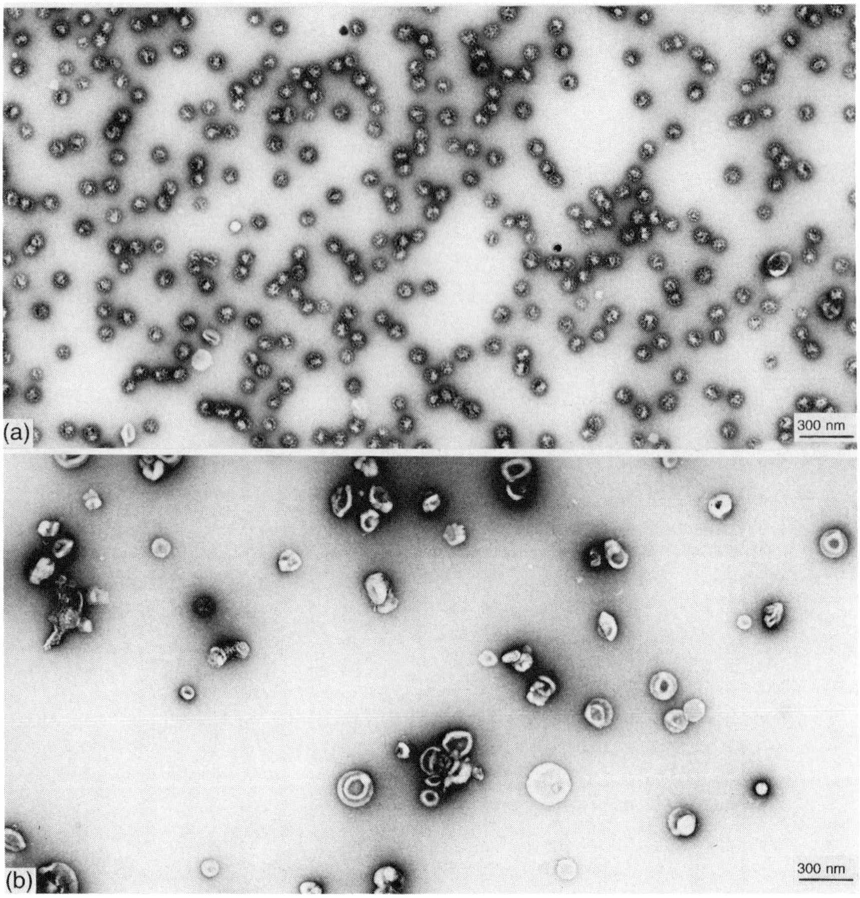

Fig. 3. Negatively stained (a) CVs and (b) smooth membranes from region II of an isopycnic Ficoll/^2H_2O gradient, prepared from an RNase-treated zucchini hypocotyl postmicrosomal supernatant. Separation has been achieved by a subsequent isopycnic centrifugation on a sucrose density gradient (see the text).

completely removed by subjecting region II to a subsequent isopycnic centrifugation in a 25% to 50% (w/w) sucrose density gradient (Depta *et al.* 1991). In such gradients the CVs band at around 40% sucrose and the contaminants equilibrate between 30% and 36% sucrose (Fig. 3 (a) and (b)).

Table 1. *Protein recoveries in coated vesicle isolations in relation to tissue pretreatment*

Source	Pretreatment[a]	Amount of start material (kg)	125 000*g*[b] pellet (mg)	CVs[c] (mg)
Zucchini	Fresh	1	456	0.51
hypocotyls	ff/ft	1	546	0.38
	sf/st	1	421	0.41
	Plasmolysed	1	204	0.42
Developing	Fresh	1	1249	5.94
pea cotyledons	sf/st	1	1895	6.54
Bovine brain	Fresh	1	770	13.32
	sf/st	1	610	12.65

Notes:
[a]ff/ft: tissue fast frozen in liquid nitrogen and fast thawed (at room temperature for 30–45 min).
Plasmolysed: tissue stirred in 0.9 M sorbitol for 1 h at 4 °C.
sf/st: tissue slowly frozen to −80 °C and slowly thawed (at 4 °C for 16 h).
[b]After prior centrifugation of homogenate at 20 000***g***.
[c]After RNase treatment, sucrose step gradient and linear Ficoll/^2H_2O gradient purification. CV protein is given as the amount of protein in region II of Ficoll/^2H_2O gradients prepared according to the method of Depta & Robinson (1986).

Recoveries

Until recently (for reviews, see Robinson *et al.* 1987; Robinson & Depta 1988) it had been held that the yields of CVs from plant sources were roughly one-tenth of those obtained from brain tissue. This statement still remains true for the materials previously studied (suspension-cultured carrot cells, bean leaves and zucchini hypocotyls). However, Harley & Beevers (1989*a*) reported achieving yields of CVs from developing pea cotyledons which were greater than those from brain tissue. We confirm here that pea cotyledons are indeed the richest source of plant CVs that we have investigated so far, although we are not in a position to substantiate the extremely high recoveries (18 mg CV per kg fresh weight) given by Harley & Beevers (1989*a*). As seen in Table 1, our yields lie around 6 mg CV per kg fresh weight of starting material.

Since we had been working constantly with fresh pea seeds and Harley & Beevers (1989*a*) used cotyledons stored at −80 °C, we looked into the possibility that freezing and thawing might somehow be responsible for this difference in yields. We therefore compared protein recoveries in CV isolations from pea cotyledons and zucchini hypocotyls which had been either fast frozen (in liquid nitrogen) then fast thawed (left to stand for 30–45 min at room temperature) or slowly frozen (placed in a refrigerator at −80 °C) and slowly thawed (overnight in a refrigerator at +4 °C) with that of fresh material. The results presented in Table 1 show that freezing and thawing does not significantly increase the yield of CVs.

The quality of postmicrosomal pellets does, nevertheless, seem to be affected by freezing/thawing. Irrespective of the speed of freezing or thawing, CVs always appear to equilibrate a little (1% to 2% sucrose equivalents) deeper in Ficoll/^2H_2O gradients than is the case with CVs from fresh material. Even more interesting is the observation that region I is almost invariably absent from Ficoll/^2H_2O gradients of slowly frozen/slowly thawed tissue (Fig. 2). This fact questions the usefulness of Ficoll/^2H_2O gradients for CV purification, especially since the purity of CVs in region II in such gradients from slowly frozen/thawed tissue tends to be higher than for region II from fresh tissue. Because of this it is possible to omit the Ficoll/^2H_2O gradient and proceed direct to an isopycnic sucrose density gradient as described above. Highly enriched CV fractions are again found at densities around 40% sucrose. It is not easy to provide an explanation for the loss of region I membranes as a result of slow freezing/thawing, but intracellular ice crystal formation could cause membranes to be fractured into larger fragments which would then sediment in the microsomal rather than postmicrosomal pellet.

Although the reason for doing it has not been given, plasmolysis has been carried out on tissues prior to homogenisation and CV isolation (Mersey *et al.* 1982; Wiedenhoeft *et al.* 1988). We have also investigated the effect of plasmolysis on CV yield, using the same conditions employed by Wiedenhoeft *et al.*, and have found that it too does not improve CV yields significantly (see Table 1).

Problems involved in CV isolation

The vacuole is the major obstacle in CV isolation from plant cells. Physically it leads to greater homogenate volumes when one is trying to isolate the corresponding amount of CVs from brain tissue. Depending upon the availability of ultracentrifuges this may result in delays in the preparation of a postmicrosomal pellet. Chemically the vacuole, because of the substances sequestered in it, can be of great potential danger to the quality of the CVs isolated. Whereas the effect of phospholipases and polyphenol oxidases can be efficiently combatted by the inclusion of bovine serum albumin and metabisulphite, respectively (Depta *et al.* 1987*b*), preventing proteolysis is a problem that we feel has not yet been adequately solved in all cases.

Inclusion of protease inhibitors such as phenylmethylsulphonylfluoride (PMSF) in the homogenisation and centrifugation media is to be recommended. Some protease inhibitors which appear to be more appropriate for plant extracts, e.g. chymostatin (Gallagher *et al.* 1986), can, for financial reasons, usually be added only at later stages in CV isolation, i.e. when volumes become smaller. Nevertheless chymostatin should be included wherever possible, even in the sample buffer used for sodium dodecyl sulphate/polyacrylamide gel electrophoresis (SDS/PAGE), since it is known to prevent effectively the 'nicking' of membrane ATPases (Addison & Scarborough 1981), which can occur during the preparation of protein samples for gel electrophoresis.

Despite these precautions we feel that many of the CV fractions which we, and others, have prepared in the past have not been adequately protected against the action of vacuolar proteases. It is known that clathrin light chains and, to a lesser extent, the assembly polypeptides are very susceptible to proteolysis (Kirchhausen & Harrison 1984; Zaremba & Keen 1985). Thus when using SDS/PAGE to compare CVs from brain tissue with those isolated from plant sources under the same conditions, one is struck by the relative weakness of bands in the plant CV lanes at positions where clathrin light chains (30–40 kDa) and the assembly polypeptides (50, 100 kDa) might be expected to be found (see appropriate Figures in Mersey *et al.* 1985; Depta & Robinson 1986; Depta *et al.* 1987*b*; Robinson & Depta 1988). Since light and heavy chains of clathrin are present with a 1 : 1 stoichiometry in the animal triskelion (Kirchhausen *et al.* 1983), and assuming this is true also for plant triskelia, the intensity of bands in SDS/PAGE corresponding to putative light-chain polypeptides in plant CVs is not what it should be.

Ironically the RNase treatment which appears to be so necesary for the preparation of plant CV fractions can be earmarked as a contributory factor in the proteolytic degradation of plant CVs. This treatment, which involves incubating the CV fractions for up to 30 min at 30 °C, may well allow some proteases to partially degrade the CVs despite the presence of PMSF. Indirectly supporting this suggestion is the observation that plant CVs which have not been treated with RNase equilibrate at densities around 50% sucrose (Mersey *et al.* 1985; A. Demmer & D. G. Robinson, unpublished observations; see also Pearse 1975, for brain CVs). In contrast, CVs that have been exposed to RNase are less dense (equilibrating at 40% sucrose, see above), suggesting that they have lost protein. Clearly it would be of considerable benefit to develop a method for CV isolation which circumvents the RNase step.

Enzymic characterisation of plant CVs

Attempts have been made to measure endomembrane and plasma membrane marker enzyme activities in CV fractions obtained from plants (Griffing *et al.* 1986; Depta *et al.* 1987*a*; Harley & Beevers 1989*b*; Depta *et al.* 1991). In no case have significant quantities of the endoplasmic reticulum marker, cytochrome *c* reductase (CCR), been detected. However, contradictory information exists with respect to markers for the Golgi apparatus and plasma membrane.

Glucan synthases

In 1986 Griffing *et al.* published a paper in which evidence was put forward for the presence of the Golgi marker, glucan synthase I, in CVs isolated from soybean protoplasts. Depta *et al.* (1987*a*, 1991) were unable to confirm this observation; instead they measured very high activities of the plasma membrane marker, glucan synthase II, in CV fractions isolated from zucchini hypocotyls. The product of this enzyme activity was shown to be a highly crystalline, microfibrillar, 1–3-β-glucan (Depta *et al.* 1987*a*).

It is clear that claims for the presence of Golgi or plasma membrane markers in CVs have to be considered in terms of the purity of the CV fractions concerned. Depta *et al.* (1991) have addressed themselves to this question. By subjecting their enriched CV fractions from zucchini hypocotyls (region II of Ficoll/^2H_2O gradients, see above) to a further isopycnic centrifugation these authors have been able to obtain CV preparations of around 95% purity (Fig. 3 (a)). In such sucrose density gradients the CVs banded around 40% sucrose and the contaminating membranes (Fig. 3 (b)) equilibrated at lower densities (30%–36% sucrose). Both glucan synthase II and naphthylphthalamic acid (NPA) binding (also a plasma membrane marker, Quail 1979) activities were found to be associated with the contaminants rather than the purified CVs. Although we have not been able to measure phosphate release in an assay for vanadate sensitive, K^+-stimulated, H^+-ATPase activity in any fraction from such sucrose density gradients, it is reasonable to assume that some of the contaminants in region II of Ficoll/^2H_2O gradients were of plasma membrane origin.

ATPases

Antibodies raised against the 70 kDa subnit of corn root tonoplast H^+-ATPase (Zimniak *et al.* 1988) cross-react with a polypeptide of molecular

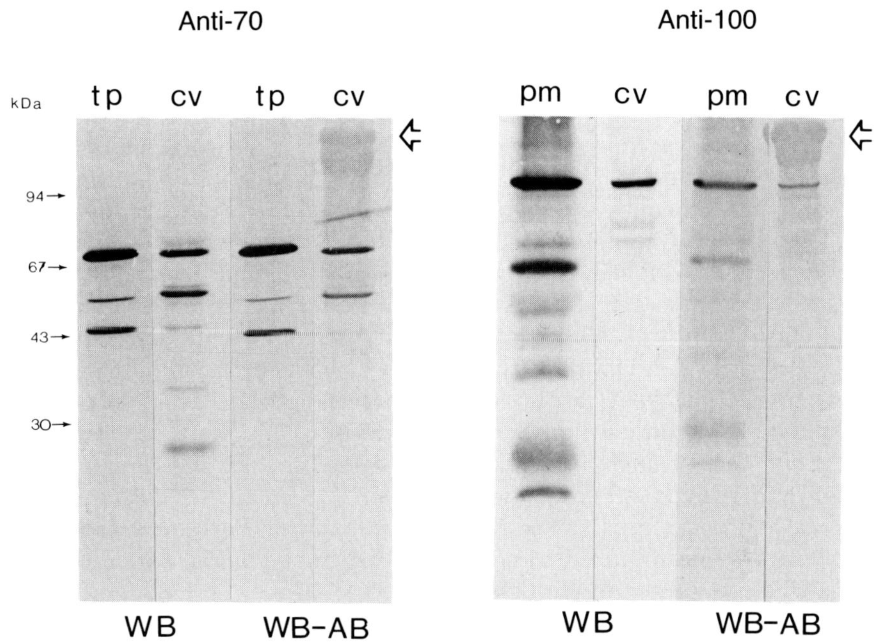

Fig. 4. Probing highly purified pea cotyledon CV fractions with antibodies against H^+-ATPase subunits (the 70 kDa subunit of the tonoplast H^+-ATPase (anti-70); the 100 kDa subunit of the plasma membrane H^+-ATPase (anti-100)). WB, Western blots (colour developed with alkaline phosphatase); WB-AB, the same Western blots stained subsequently with amido black; tp, purified tonoplast from sugar beet; pm, phase partioning purified plasma membrane from zucchini hypocotyls; cv, CV fractions; large arrow indicates the position of clathrin heavy chain at 190 kDa. Amount of protein in cv lanes 14 (18) μg; in tp lane 7.5 μg; in pm lane 18 μg.

mass around 70 kDa in Western blots of plant CV preparations (Fichmann *et al.* 1989; Depta *et al.* 1991; see also Fig. 4). Although we have not yet been able to demonstrate proton-pumping in highly purified CV fractions from plant sources, these immunological results suggest strongly that plant CVs are similar to their animal counterparts in possessing a vacuolar (V)-type H^+-ATPase (Nelson 1988; Forgac 1989). This is further strengthened by the observation that the tonoplast H^+-ATPase antibodies mentioned above also cross-react with the corresponding subunits in coated vesicles isolated from bovine brain (Fichmann *et al.* 1989).

We have probed highly purified (and therefore presumably free from plasma membrane contaminants) CV fractions isolated from pea cotyledons (Fig. 4) and zucchini hypocotyls (see Depta *et al.* 1991), with antibodies raised against the 100 kDa subunit of corn root plasma membrane H^+-ATPase. Whereas this antibody in the hands of Fichmann *et al.* (1989) apparently did not recognise any polypeptide in Western blots prepared from zucchini CVs we have been able to show a cross-reaction with a polypeptide of molecular mass around 100 kDa for CVs from both tissue types. In contrast to most other V-type H^+-ATPases, brain CVs and chromaffin granules also possess a subunit of around 110 kDa (Percy & Apps 1986; Arai *et al.* 1987). Until recently it has not been clear what purpose this large subunit serves. It now seems that, in both brain CVs and chromaffin granules, this polypeptide is part of an associated, but different, ATPase. Unlike V-type ATPases, which are hetero-oligomeric complexes of 530 kDa, and are insensitive to vanadate (Forgac 1989), the 110 kDa polypeptides appear to be dimeric and are vanadate sensitive (Moriyama & Nelson 1988; Xie *et al.* 1989). They are thus much more akin to E_1E_2-type ATPases present in, for example, the plasma membrane. Although it is not yet known whether there is any antigenic similarity between the 110 kDa CV and E_1E_2 subunits, our recent results might be interpreted in the light of this possibility.

Luminal contents of plant CVs

CVs in animal cells are usually regarded as being the carriers of specific ligands ('cargo', Pearse & Bretscher 1981). Immunologically different classes of assembly proteins ('adaptors' *sensu* Pearse & Bretscher 1981) apparently enable the cell to distinguish between CVs involved in endocytic and lysosomal protein transport. The literature on animal CVs contains numerous examples for the detection of extracellular or endomembrane (lysosomal)-type ligands in isolated CVs (for reviews, see Goldstein *et al.* 1985; von Figura & Hasilik, 1986). Comparable data for plant CVs have only recently become available, and this pertains, as yet, only to the transport of vacuolar proteins.

As often happens analogous situations in animal cells serve as experimental models for plant cell biologists. Thus the fact that the plant vacuole is often regarded as being the lysosomal compartment of plant cells (see, for example, Matile 1978), taken together with the observation that membrane structures at the *trans* pole of the Golgi apparatus in both plant and animal cells have a certain morphological similarity (Hillmer *et al.* 1988), forces one to pose the question: are CVs which are present in the *trans*-Golgi network of plant cells involved in vacuolar protein transport? We believe that this question can be answered in the affirmative, although immunocytochemical data are not yet available to prove that the CVs in question really are Golgi derived.

Cotyledon cells of maturing legume seeds have proved to be excellent objects for investigations into this problem, since these cells are geared up for vacuolar protein synthesis, and the proteins in question have been well characterised (Higgins 1984). Working independently of one another, Harley & Beevers (1989*b*) and Robinson *et al.* (1989) have isolated CVs from developing pea cotyledons and have demonstrated immunologically the presence of vacuolar polypeptides in them. Harley & Beevers (1989*b*) were able to detect the 23 kDa precursor polypeptide of pea lectin, and Robinson *et al.* (1989) the 60 kDa precursor of legumin in pea CV preparations. Two lines of evidence indicate that these polypeptides are sequestered within the CV lumen. Firstly, only when legumin becomes deposited in the protein bodies (which are derived from the vacuoles through fission) is it cleaved into 40 and 20 kDa subunits which are then held together by a disulphide bridge. If therefore the latter polypeptides were detected by legumin antibodies in Western blots of SDS/PAGE run under reducing conditions, this would mean that CV-associated legumin came from ruptured protein bodies and is artefactually associated with the surface of the CVs. This is, however, not the case. Secondly, Harley & Beevers (1989*b*) have shown that the degradation of pea lectin through exogenously added proteinase K can occur only in the presence of detergents.

The CV preparations employed by Harley & Beevers (1989*b*) and Robinson *et al.* (1989) were in fact the pelleted region II membranes of Ficoll/^2H_2O gradients. As discussed above, such CV fractions are contaminated by up to 20% with endomembrane vesicles or fragments. It is therefore of paramount importance that these should be ruled out as the possible carriers of the polypeptide antigens just described. However, highly purified CV fractions from pea cotyledons (see above and Fig. 5) still react positively when challenged with legumin antibodies; they also show a positive reaction towards antibodies directed against vicilin, the other major storage protein in pea cotyledons (Fig. 6).

Additional evidence in support of a role for CVs in vacuolar protein transport has been provided by Harley & Beevers (1989*b*). These authors were able to demonstrate the presence, in pea cotyledon CVs, of several hydrolases (α-galactosidase, α-mannosidase, β-hexosaminidase), which are synthesised and become deposited during seed development in legumes (see e.g. Neely & Beevers 1980). According to Harley & Beevers (1989*b*) the endoplasmic reticulum and Golgi apparatus markers CCR and IDPase were virtually absent from the pea cotyledon CVs used

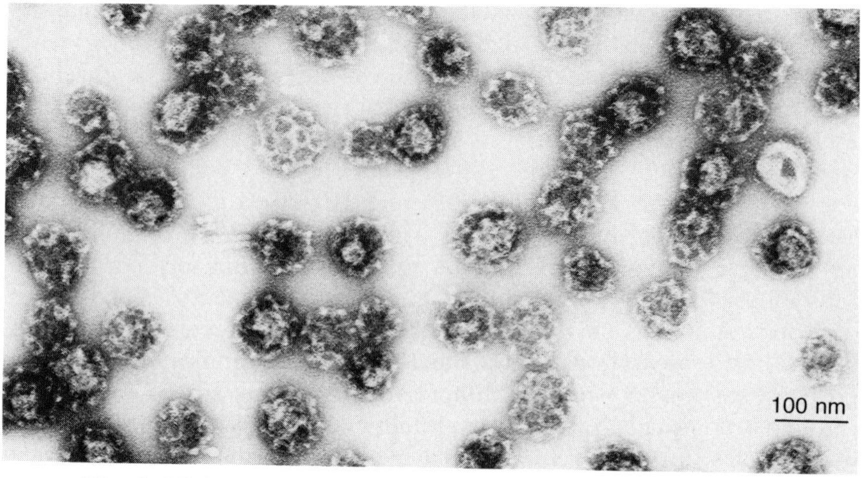

Fig. 5. Highly purified CVs isolated from developing pea cotyledons. Negatively stained preparation.

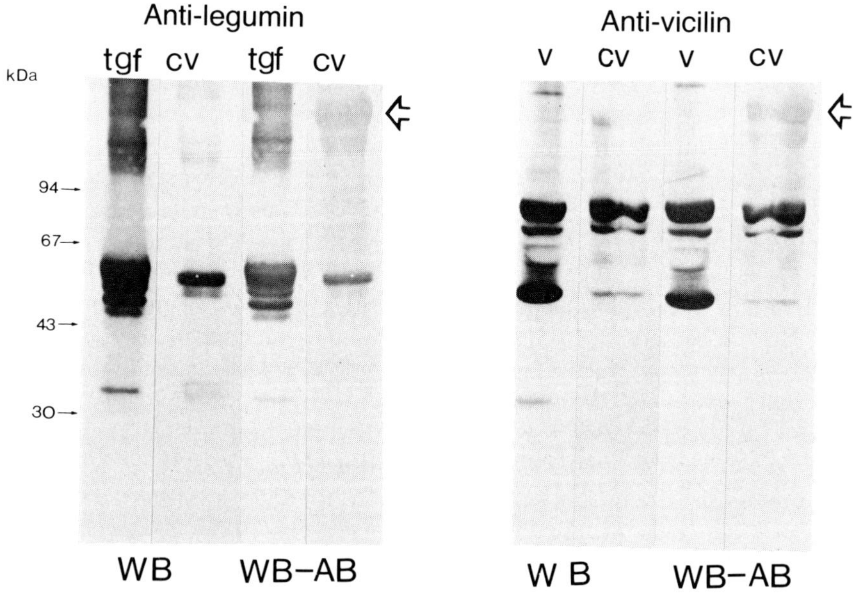

Fig. 6. Probing highly purified pea cotyledon CV fractions with monospecific polyclonal antibodies raised against the major storage proteins legumin and vicilin. WB, Western blot; WB-AB, the same subsequently stained with amido black; tgf, total globulin fraction from pea cotyledons; v, pea cotyledon vicilin standard; cv, CV fractions; the arrow points to the clathrin heavy chain at 190 kDa. Amount of protein in cv lanes 15 μg, in tgf lane 10 μg.

in their study. Whilst we, too, cannot measure CCR activity, in our hands detergent-released IDPase activity can still be measured in pea cotyledon CV preparations (B. Hoh & D. G. Robinson, unpublished observations). Since these preparations are highly purified we can rule out Golgi apparatus contamination as being responsible for this activity. Instead we consider the activity to be due to a vacuole-destined phosphatase which can utilise IDP as a substrate.

Acknowledgements

We thank Heike Freundt for help in preparing the manuscript and Bernd Rauffeisen for the illustrations. Drs R. R. D. Croy and J. A. Gatehouse (University of Durham), L. Taiz and R. T. Leonard (University of California) are gratefully thanked for supplying us with antibodies. Last, but not least, we acknowledge the generous financial support of the Deutsche Forschungsgemeinschaft.

References

Addison, R. & Scarborough, G. A. (1981). Solubilization and purification of the *Neurospora* plasma membrane H$^+$-ATPase. *Journal of Biological Chemistry*, **256**, 13 165–71.

Arai, H., Berne, M., Terres, G., Terres, H., Pupolo, K. & Forgac, M. (1987). Subunit composition and ATP site labelling of the coated vesicle proton-translocating adenosine triphosphatase. *Biochemistry*, **26**, 6632–8.

Campbell, C. H., Fine, R. E., Squicciarini, J. & Rome, L. H. (1983). Coated vesicles from rat liver and calf brain contain cryptic mannose 6-phosphate receptors. *Journal of Biological Chemistry*, **258**, 2628–33.

Coleman, J., Evans, D., Hawes, C., Horsley, D. & Cole, L. (1987). Structure and molecular organization of higher plant coated vesicles. *Journal of Cell Science*, **88**, 35–45.

Depta, H., Andreae, M., Blaschek, W. & Robinson, D. G. (1987*a*). Glucan synthase II activity in a coated vesicle fraction from zucchini hypocotyls. *European Journal of Cell Biology*, **45**, 219–23.

Depta, H., Freundt, H., Hartmann, D. & Robinson, D. G. (1987*b*). Preparation of a homogeneous coated vesicle fraction from bean leaves. *Protoplasma*, **136**, 154–60.

Depta, H. & Robinson, D. G. (1986). The isolation and enrichment of coated vesicles from suspension-cultured carrot cells. *Protoplasma*, **130**, 162–70.

Depta, H., Robinson, D. G., Holstein, S. E. H., Lützelschwab, M. & Michalke, W. (1991). Membrane markers in highly purified clathrin coated vesicles from *Cucurbita* hypocotyls. *Planta* **183**, 434–42.

Fichmann, J., Taiz, L., Gallagher, S., Leonard, R. T., Depta, H. &

Robinson, D.G. (1989). Immunochemical comparison of the coated vesicle H^+-ATPase of plants and animals. *Protoplasma*, **153**, 117–25.

Forgac, M. (1989). Structure and function of vacuolar class of ATP-driven proton pumps. *Physiological Reviews*, **69**, 765–96.

Gallagher, S.R., Carroll, E.J., Jr & Leonard, R.T. (1986). A sensitive diffusion plate assay for screening inhibitors of protease activity in plant cell fractions. *Plant Physiology*, **81**, 869–74.

Goldstein, J.L., Brown, M.S., Anderson, R.G.W., Russell, D.W. & Schneider, W.J. (1985). Receptor-mediated endocytosis: concepts emerging from the LDL receptor system. *Annual Review of Cell Biology*, **1**, 1–39.

Griffing, L.R., Mersey, B.G. & Fowke, L.C. (1986). Cell-fractionation analysis of glucan synthase I and II distribution and polysaccharide secretion in soybean protoplasts. *Planta*, **167**, 175–82.

Harley, S.M. & Beevers, L. (1989*a*). Isolation and partial characterization of clathrin-coated vesicles from pea (*Pisum sativum* L.) cotyledons. *Protoplasma*, **150**, 103–9.

Harley, S.M. & Beevers, L. (1989*b*). Coated vesicles are involved in the transport of storage proteins during seed development in *Pisum sativum* L. *Plant Physiology*, **91**, 674–8.

Higgins, T.J.V. (1984). Synthesis and regulation of major proteins in seeds. *Annual Review of Plant Physiology*, **35**, 191–221.

Hillmer, S., Freundt, H. & Robinson, D.G. (1988). The partially coated reticulum and its relationship to the Golgi apparatus in higher plant cells. *European Journal of Cell Biology*, **47**, 206–12.

Keen, J.H., Willingham, M.C. & Pastan, I.H. (1979). Clathrin-coated vesicles: isolation, dissociation and factor-dependent reassociation of clathrin baskets. *Cell*, **16**, 303–12.

Kirchhausen, T. & Harrison, S.C. (1984). Structural domains of heavy clathrin chains. *Journal of Cell Biology*, **99**, 1725–34.

Kirchhausen, T., Harrison, S.C., Parham, P. & Brodsky, F.M. (1983). Location and distribution of the light chains in clathrin trimers. *Proceedings of the National Academy of Sciences, USA*, **80**, 2481–5.

Matile, P. (1978). Biochemistry and function of vacuoles. *Annual Review of Plant Physiology*, **29**, 193–213.

Mersey, B.G., Fowke, L.C., Constabel, F. & Newcomb, E.H. (1982). Preparation of a coated vesicle-enriched fraction from plant cells. *Experimental Cell Research*, **141**, 459–63.

Mersey, B.G., Griffing, L.G., Rennie, P.J. & Fowke, L.C. (1985). The isolation of coated vesicles from protoplasts of soybean. *Planta*, **163**, 317–27.

Moriyama, Y. & Nelson, N. (1988). Nucleotide binding sites and chemical modification of the chromaffin granule proton ATPase. *Journal of Biological Chemistry*, **263**, 14 723–9.

Neely, R.S. & Beevers, L. (1980). Glycosidases from cotyledons of *Pisum sativum* L. *Journal of Experimental Botany*, **31**, 299–312.

Nelson, N. (1988). Structure, function and evolution of protein-ATPases. *Plant Physiology*, **86**, 1–3.
Pearse, B.M.F. (1975). Coated vesicles from pig brain: purification and biochemical characterization. *Journal of Molecular Biology*, **97**, 93–8.
Pearse, B.M.F. (1982). Coated vesicles from human placenta carry ferritin, transferrin, and immunoglobulin G. *Proceedings of the National Academy of Sciences, USA*, **79**, 451–5.
Pearse, B.M.F. & Bretscher, M.S. (1981). Membrane recycling by coated vesicles. *Annual Review of Biochemistry*, **50**, 85–101.
Percy, J.M. & Apps, D.K. (1986). Proton-translocating ATPase of chromaffin granule membranes. *Biochemical Journal*, **239**, 77–81.
Quail, P.H. (1979). Plant cell fractionation. *Annual Review of Plant Physiology*, **300**, 425–84.
Robinson, D.G., Andreae, M., Depta, H., Hartmann, D. & Hillmer, S. (1987). Coated vesicles in plants: characterization and function. In *Plant Membranes. Structure, Function, Biogenesis*, ed. C. Leaver & H. Sze, UCLA Symposia on Molecular and Cellular Biology, vol. 63, pp. 341–58. New York: Alan R. Liss.
Robinson, D.G., Balusek, K. & Freundt, H. (1989). Legumin antibodies recognize polypeptides in coated vesicles from developing pea cotyledons. *Protoplasma*, **150**, 79–82.
Robinson, D.G. & Depta, H. (1988). Coated vesicles. *Annual Review of Plant Physiology and Molecular Biology*, **39**, 53–99.
von Figura, K. & Hasilik, A. (1986). Lysosomal enzymes and their receptors. *Annual Review of Biochemistry*, **55**, 167–93.
Wiedenhoeft, R.E., Schmidt, G.W. & Palevitz, B.A. (1988). Dissociation and reassembly of soybean clathrin. *Plant Physiology*, **86**, 412–16.
Xie, X.-S., Stone, D.K. & Racker, E. (1989). Purification of a vanadate-sensitive ATPase from clathrin-coated vesicles of bovine brain. *Journal of Biological Chemistry*, **264**, 1710–14.
Zaremba, S. & Keen, J.H. (1985). Limited proteolytic digestion of coated vesicle assembly polypeptides abolishes reassembly activity. *Journal of Cellular Biochemistry*, **28**, 47–58.
Zimniak, L., Dittrich, P., Gogarten, J.P., Kibak, H. & Taiz, L. (1988). The cDNA sequence of the 69 kDa subunit of the carrot vacuolar H^+-ATPase: homology to the beta-chain of F_0F_1-ATPases. *Journal of Biological Chemistry*, **263**, 9102–12.

K. J. OPARKA, L. COLE, K. M. WRIGHT, C. R. HAWES, D. E. EVANS AND J. O. D. COLEMAN

Fluid-phase endocytosis and the subcellular distribution of fluorescent probes in plant cells

Endocytosis is a collective term which is used to describe the mechanisms by which eukaryotic cells ingest selected regions of their plasma membranes. Typically, endocytosis begins with an invagination of the plasma membrane, which subsequently leads to the formation of a closed vesicle that becomes detached from the parent membrane. In the cytoplasm, the endocytotic vesicle fuses with the first compartment of a complex membrane system that makes up the endocytotic pathway. This pathway of membrane traffic serves as a discontinuous channel of communication between the extracellular environment and intracellular organelles.

Endocytotic vesicles consist of membrane components (protein and lipid) which are derived from the excised region of the plasma membrane as well as material which binds (specifically and non-specifically) to the extracytoplasmic face of the plasma membrane. In addition to these molecules, solutes dispersed in the fluid phase of the extracellular medium are inevitably trapped in the lumen of the vesicle during the process of engulfment. This fluid entrapment and internalisation is known as fluid-phase endocytosis (formerly referred to as pinocytosis). Endocytosis is never found in prokaryotes, even in those with large cells, and it is often cited as a characteristic feature of eukaryotes. The extent to which this feature is conserved in walled eukaryotic cells (plants and fungi) has in the past been a matter of some controversy. A wide range of compounds have now been shown to be taken up into plant cells and protoplasts by endocytosis, including proteins such as ferritin, colloidal gold conjugates and heavy-metal salts (for reviews, see Coleman *et al.* 1988; Robinson & Hillmer 1990; Fowke *et al.*, this volume).

In receptor-mediated endocytosis animal cells select and concentrate specific molecules (ligands) from the surrounding medium and exclude others. Ligands bind with high affinity (equilibrium binding constants of 10^{7}–10^{10} M^{-1}) to *trans*-plasma membrane proteins (receptors) which are either already located in clathrin-coated pits or are recruited into coated pits after ligand binding. Sequence information in the cytoplasmic tails of the receptor proteins is responsible for the interaction between the receptor–ligand complex and the clathrin triskelion. This interaction is mediated by accessory coat proteins known as adaptor proteins (see Coleman *et al.*, this volume).

By generating a high surface area to volume ratio in the formation of coated vesicles, the volume of external fluid trapped in the lumen of the vesicle is kept as small as possible, thus ensuring minimal uptake of unwanted molecules. For example a coated vesicle with a diameter of 100 nm would have a volume of approximately 5×10^{-16} ml and a surface area of 3×10^{-10} cm^2. In other words vesicles with a total volume of 1 ml would generate a surface area of approximately 7.5 m^2! In short, receptor-mediated endocytosis is a concentrative process of high biological specificity and is clathrin dependent. In contrast, fluid-phase endocytosis is non-specific and cannot concentrate ligands.

Despite the specificity of receptor-mediated endocytosis, it is still a controversial matter whether fluid-phase endocytosis is clathrin dependent, clathrin independent or both (for a review, see van Deurs *et al.* 1989). There is evidence to support the contention that there is a clathrin-independent fluid-phase pathway in the cells of organisms ranging from amoebae to the higher vertebrates and that different cell types may exploit fluid-phase and receptor-mediated endocytosis in different ways.

In plant cells, although clathrin-mediated endocytosis of non-physiological ligands has been observed (see Coleman *et al.*, this volume), receptor-mediated endocytosis of a physiological ligand has so far not been demonstrated (however, see Horn *et al.* 1989). In addition, studies of fluid-phase endocytosis have not distinguished between clathrin-dependent and clathrin-independent uptake. The presence and expression of a clathrin-independent pathway in plant cells raises a number of challenging questions. For example: is uptake via this pathway initiated by physical changes of the plasma membrane which are imposed by the addition of membrane during secretion? As the entrapment process is non-selective, is there some post-internalisation editing mechanism which sorts wanted from unwanted molecules? What are the destinations of these molecules? What are the osmotic implications which result from this route of uptake? Before these questions can be addressed, fluid-phase endocytosis in plant cells must be demonstrated satisfactorily using reliable experimental techniques. Accordingly, one of the chief objectives of this chapter will be to discuss the techniques employed to follow fluid-phase endocytosis in plants (in particular, fluorescent methods), to point out the problems and pitfalls which can occur and to consider what practical conclusions can be safely drawn.

The experimental approach

Identification of fluid-phase endocytosis

One of the major problems encountered in the study of endocytosis, particularly at the level of the light microscope, arises from trying to distinguish whether or not a molecule has entered by a fluid-phase or a receptor-mediated pathway. A widely used indicator is the kinetics of uptake. Fluid-phase endocytosis shows linear uptake kinetics with respect to increasing substrate concentration (an inevitable consequence of the non-selective uptake of external fluid) and shows no evidence of saturation over a large (mg ml^{-1}) external solute range. Uptake of labelled solute is not reduced when unlabelled solute is added to the medium, and binding to the cell surface is not detectable at either 4 °C or 37 °C. Furthermore, when more than one solute is tested, the uptake of each occurs at a similar rate. By contrast, uptake by receptor-mediated endocytosis is saturable with respect to substrate concentration (a consequence of the saturability of receptor binding sites at high external ligand concentrations). However, high rates of receptor-mediated endocytosis can occur for hours, even with saturating levels of ligand (Steinman *et al.* 1983). Other complications may also occur; for example, Lang & de Chastellier (1985) found that macrophages internalise horseradish peroxidase (a commonly used fluid-phase marker) by both fluid-phase endocytosis and binding to mannose receptors.

A useful way of distinguishing clathrin-independent fluid phase uptake from the coated pit/coated vesicle pathway is to employ agents which selectively inhibit the clathrin-mediated pathway. This can be achieved mainly by three methods: (1) hypertonicity, (2) cytosol acidification and (3) K^+ depletion.

Hypertonicity and K^+ depletion

In leukocytes (Daukas & Zigmond 1985) and hepatocytes (Oka & Weigel 1987), hypertonic media inhibited receptor-mediated endocytosis but not fluid-phase endocytosis. Potassium depletion combined with hypotonic shock strongly inhibits receptor-mediated endocytosis in some cell types (Larkin *et al.* 1986). However, this effect is not universal as some cells are not affected, whereas others are very sensitive. Both hypertonicity and K^+ depletion appear to affect receptor-mediated endocytosis by either blocking or perturbing clathrin assembly at the plasma membrane (Heuser & Anderson 1989; Madshus *et al.* 1987).

Acidification

Cytosolic acidification, induced by NH_4Cl prepulsing, pH clamping with valinomycin and nigericin or by the uptake of undissociated acetic acid, blocks receptor-mediated endocytosis (Sandvig *et al.* 1987, 1989). In acidified cells, the coated pits appear to be immobilised at the plasma membrane and the process of 'pinching off' is inhibited.

There have been very few studies of fluid-phase endocytosis in plants and consequently the full range of treatments which might influence this process is as yet unclear.

Probes for fluid-phase endocytosis in plants

For studies of fluid-phase endocytosis, the most convenient substrates to use are marker molecules which by virtue of their size or charge cannot enter cells other than by endocytosis.

Macromolecules

Macromolecules whose uptake can be followed by a specific label or by a specific reaction (e.g. enzymes), and whose intracellular localisation can be visualised by light or electron microscopy, are ideal marker substrates. Fluorescently labelled globular proteins such as bovine serum albumin (BSA) or the enzyme horseradish peroxidase (HRP) are commonly used as fluid-phase markers in studies of animal cells (see Swanson *et al.* 1985).

In plant tissues the cell wall can restrict the use of macromolecular markers. For example Baron-Epel *et al.* (1988) found that, in soybean cells, macromolecules with Stokes' radii $\leqslant$3.3 nm could pass freely through the cell wall. Those with Stokes' radii of 3.3–4.6 nm were severely hindered and molecules with Stokes' radii >4.6 could not penetrate the cell wall. Although some globular proteins are not restricted by their size from entering the cell wall, their net ionic charge at physiological pH values may result in large scale binding to the cell wall. For example, horse-heart cytochrome *c* (M_r 12 000), which might be expected to permeate the cell wall, binds strongly to the cell walls of cultured carrot cells. Modification of the protein by acetylation of lysine residues prevents this binding (Chalmers 1988).

The molecular sieving properties of the cell wall can be circumvented experimentally by using plant protoplasts. However, allowances must be made for the possibility that the endocytotic behaviour of protoplasts may differ significantly from that of intact cells (Cole *et al.* 1990).

Fluorescent molecules are good fluid-phase markers as their uptake can be determined quantitatively by spectrofluorimetry, their intracellular localisation by fluorescence microscopy (Fig. 2) and the dynamics of their intracellular compartmentation by confocal laser microscopy (Fig. 3).

Fluorescent marker-labelled dextrans are convenient tracer macromolecules for fluid-phase studies in plant cells. These molecules can be prepared by conjugating the fluorescent dye fluorescein-5-isothiocyanate (FITC) to dextrans (glucose homopolymers) (see Fig. 1 (a)). Fluorescent dextrans are water-soluble molecules with low toxicity and a net anionic charge at pH 7.0. By selecting a dextran with a suitable chain length one can obtain a marker molecule with a Stokes' radius small enough to allow free permeation through the cell wall. For example dextrans with molecular weights up to 20 000 (Stokes' radii 3.3 nm) can pass through the cell wall unhindered (Baron-Epel *et al.* 1988; Cole *et al.* 1990).

Despite their attractiveness as fluorescent markers, the use of FITC-dextrans to study endocytosis in plant cells must be carried out with great care. Doubt has been cast on earlier reports (Makarow 1985; Makarow & Nevalainen 1987) that yeast cells take up FITC-dextrans by fluid-phase endocytosis and concentrate the molecules in their vacuoles. Preston *et al.* (1987) demonstrated that commercial preparations of FITC-dextran are highly contaminated with low molecular weight fluorescent compounds (e.g. unconjugated FITC) which enter the cells by a non-endocytotic route and are sequestered in the vacuoles. After purification of the dextrans no evidence for endocytosis was obtained.

A recent study of endocytosis in suspension-cultured plant cells (Cole *et al.* 1990) indicated the presence of low molecular weight fluorescent compounds (probably FITC), in commercial preparations of FITC-dextrans, which varied from batch to batch. In experiments with unpurified FITC-dextran, these contaminants entered the cell by direct penetration of the plasma membrane and were sequestered exclusively in a pleiomorphic vacuolar system identical with that loaded with dye when cells were incubated with FITC (Figs. 2 (c) to (f) and 3). This pattern of distribution could be misinterpreted as fluid-phase endocytosis of the FITC-dextran. It should be noted, however, that contaminant-free FITC-dextran (M_r 20 000), which can readily permeate the cell wall, labelled the vacuolar system of intact carrot cells (see Fig. 2 (a) and (b)).

Low molecular weight fluorescent probes

There are a number of fluorescent, low molecular weight polar compounds which, because of their ionic charge, are membrane impermeant and therefore useful as endocytotic markers. These compounds are

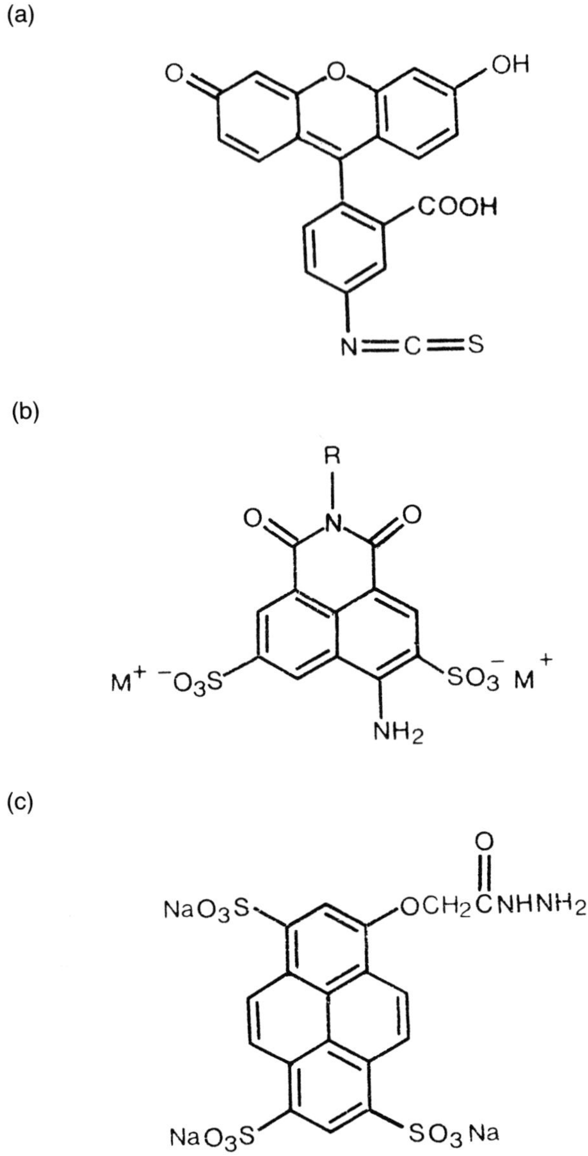

Fig. 1. Structural formulae of fluorescent probes. (a) Fluorescein-5-isothiocyanate (FITC). This is linked via the thiocyanate group to amino dextrans or to free amino groups in proteins and peptides. (b) Lucifer dyes. For LY-CH, R is -NH-CO-NH-NH_2; for LY-VS, R is *M*-phenyl-SO_2-CH=CH_2. M is Li, K or NH_4. (c) Cascade Blue hydrazide.

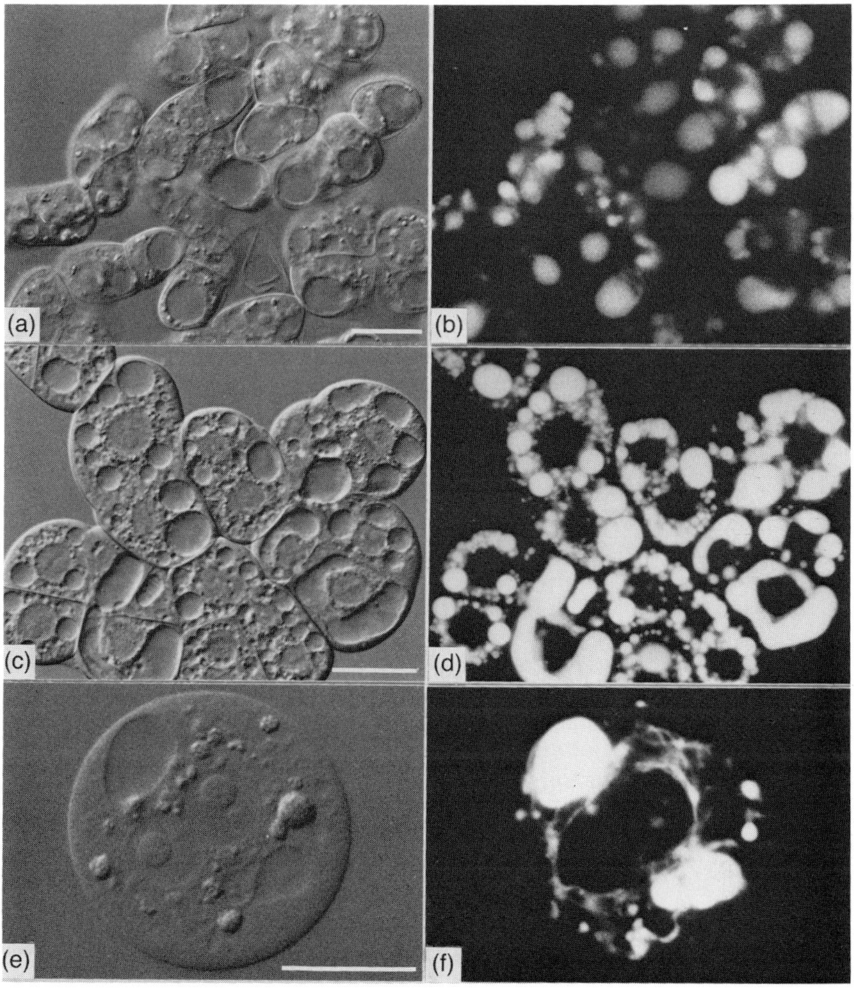

Fig. 2. Uptake of fluorescent probes into carrot cells. (a) and (b) Differential interference contrast and epifluorescence micrographs of suspension-cultured carrot cells incubated with purified (3×) 20 kDa FITC-dextran (12.5 mg ml^{-1}, pH 5.6) for 30 min. at 25 °C. Fluorescence accumulates only in the vacuoles.

(c) and (d) Uptake of FITC (0.05 mg ml^{-1}, pH 6.5) into the vacuoles of suspension-cultured carrot cells after 5 min of incubation in the dye.

(e) and (f) Uptake of FITC (0.05 mg ml^{-1}, pH 5.0) into the vacuolar system of protoplasts from suspension-cultured carrot cells after 5 min of incubation in the dye. Note that the vacuolar system is in part tubular. The bars represent 20 μm.

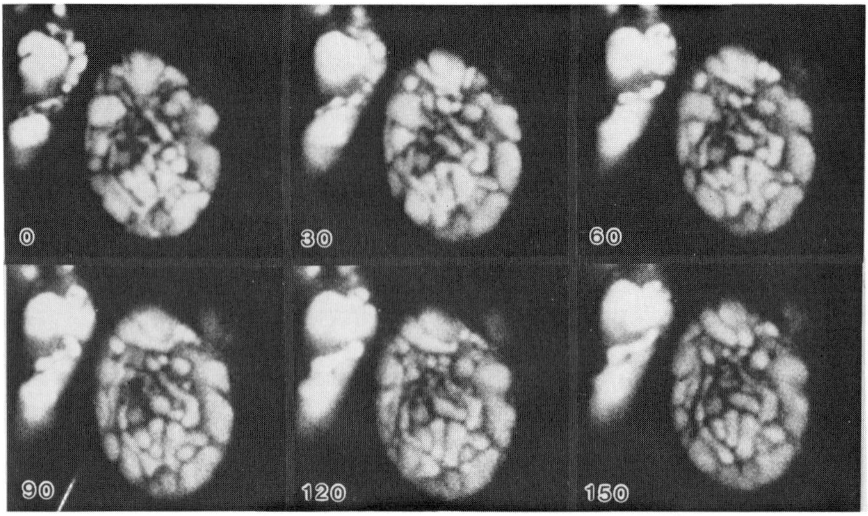

Fig. 3. Time-lapse laser scanning confocal microscope series of carrot cells loaded with FITC as in Fig. 2 (d) and recorded at 30 s intervals. These micrographs demonstrate the pleiomorphic nature of the vacuolar system near the cortex of the cell. There is considerable budding, fusion and movement of the vacuoles.

potentially useful for studies of fluid-phase endocytosis in plant cells as they are small enough to pass through the cell wall.

Lucifer dyes

The Lucifer dyes, Lucifer Yellow carbohydrazide (LY-CH) and Lucifer Yellow vinyl sulphone (LY-VS) are intensely fluorescent 4-aminonaphthalimides which are frequently used as biological tracers. The dyes differ in the substituent on the imide nitrogen of the aminonaphthalimide moiety (see Fig. 1 (b)). The dyes have similar spectral properties with absorption maxima at 280 and 430 nm and emission maxima at 540 nm (Stewart 1978, 1981), but differ in their chemical properties. Lucifer Yellow VS (LY-VS) reacts with amino and sulphydryl groups and as a result combines rapidly and covalently with proteins under mild conditions. Lucifer Yellow CH (LY-CH) is less reactive, although the hydrazido group can react with aliphatic aldehydes at room temperature. LY-VS is used less frequently than the LY-CH form which has a number of distinct properties that make it an exceptionally useful tool for studies of fluid-phase endocytosis (Table 1).

Table 1. *Properties of Lucifer Yellow CH (after Stewart 1978)*

1.	Highly soluble in water
2.	High quantum yield (detectable at 10^{-6} M)
3.	Quantum yield stable between pH 1 and 10
4.	Lipid insoluble. Partition coefficient $<7\times10^{-11}$ (toluene : buffer)
5.	Highly ionised. p*K* values of disulphonates <0.7
6.	Reacts with aldehyde fixatives
7.	Retains fluorescence following resin embedding
8.	Low toxicity

Of the properties listed in Table 1, the very low partition coefficient and low pK_a values of the disulphonate groups render the molecule membrane impermeant. For studies of fluid-phase endocytosis it is absolutely imperative that a probe should be unable to diffuse across the plasma membrane. Other commonly used fluorochromes, e.g. fluorescein, do not meet this requirement fully, being partially undissociated in the region of pH 2–6, thereby allowing diffusion of the neutral form of the molecule across membranes. Miller *et al.* (1983) appear to have been the first to utilise the membrane-impermeant properties of LY-CH in a study of fluid-phase endocytosis in macrophages. Subsequently, Swanson *et al.* (1985) made a comprehensive study of the uptake kinetics of LY-CH compared with other fluid-phase markers and concluded that it was an 'excellent probe for pinocytosis [endocytosis]'. The dye has now been used extensively in studies of fluid-phase endocytosis in animal cells (see e.g. Buckmaster *et al.* 1987; Sandvig *et al.* 1987; Swanson *et al.* 1987; Swanson 1989). Riezman (1985) also used LY-CH to study fluid-phase endocytosis in yeast, finding that the dye was taken up in a characteristic time-, temperature- and energy-dependent manner. The inability of yeast mutants, defective in endocytosis, to take up the dye (Riezman 1985; Riezman *et al.* 1986) is further evidence in favour of its reliability as a marker for fluid-phase endocytosis.

Uptake of Lucifer Yellow CH into intact plant tissues

Oparka & Prior (1988) were the first to report evidence for the endocytotic uptake of LY-CH into the vacuoles of potato-tuber storage parenchyma. The dye, in addition to moving freely in the apoplast, was sequestered in the central vacuole and could be visualised following the fixation and embedding of the tissue. Clathrin-coated vesicles were found to be abundant in the cytoplasm of the storage cells, although their involvement in endocytosis could not be demonstrated.

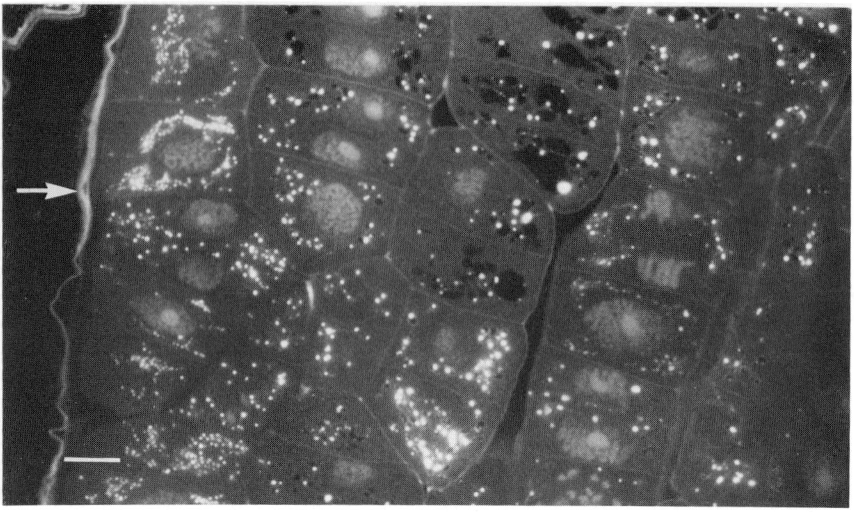

Fig. 4. Internalisation of LY-CH into roots. Longitudinal section of an intact barley (*Hordeum vulgare* L. cv. Heriot) root tip which had been submerged in 1% (w/v) aqueous LY-CH for 2 h. Accumulation of the dye is apparent in provacuoles in the outer cells of the root cortex. Cells located deeper in the cortex have larger vacuoles but deposits of the dye are still present within them. The arrow denotes the external surface of the root. The bar represents 10 μm. (From Oparka *et al.* 1988.)

In subsequent experiments LY-CH was applied to the apical 1 cm of intact barley root tips. The dye moved rapidly through the apoplast and accumulated in the entire vacuolar system of the root tip (Fig. 4). When LY-CH was applied to root tips in which the Casparian band was undeveloped, the dye moved freely in the apoplast and, in addition, entered small vacuoles in cells of both the cortex and stele (Fig. 5 (a)). In a second experiment, when LY-CH was applied to a 1 cm zone approximately 6 cm behind the root tip, the fully developed Casparian band of

Fig. 5. Internalisation of LY-CH into roots. (a) Transverse section of an intact barley root taken through the region of vascular differentiation. Note the presence of mature protophloem sieve elements (P). The LY-CH has penetrated the entire apoplast of the root. In addition, deposits of the dye are present in cells of both the cortex and stele (arrowheads). The endodermis (E) is identifiable but the Casparian band has not yet formed on the radial walls. The bar represents 10 μm. (From Oparka *et al.* 1988.)

(b) Transverse section of a barley root to which 1% LY-CH was applied

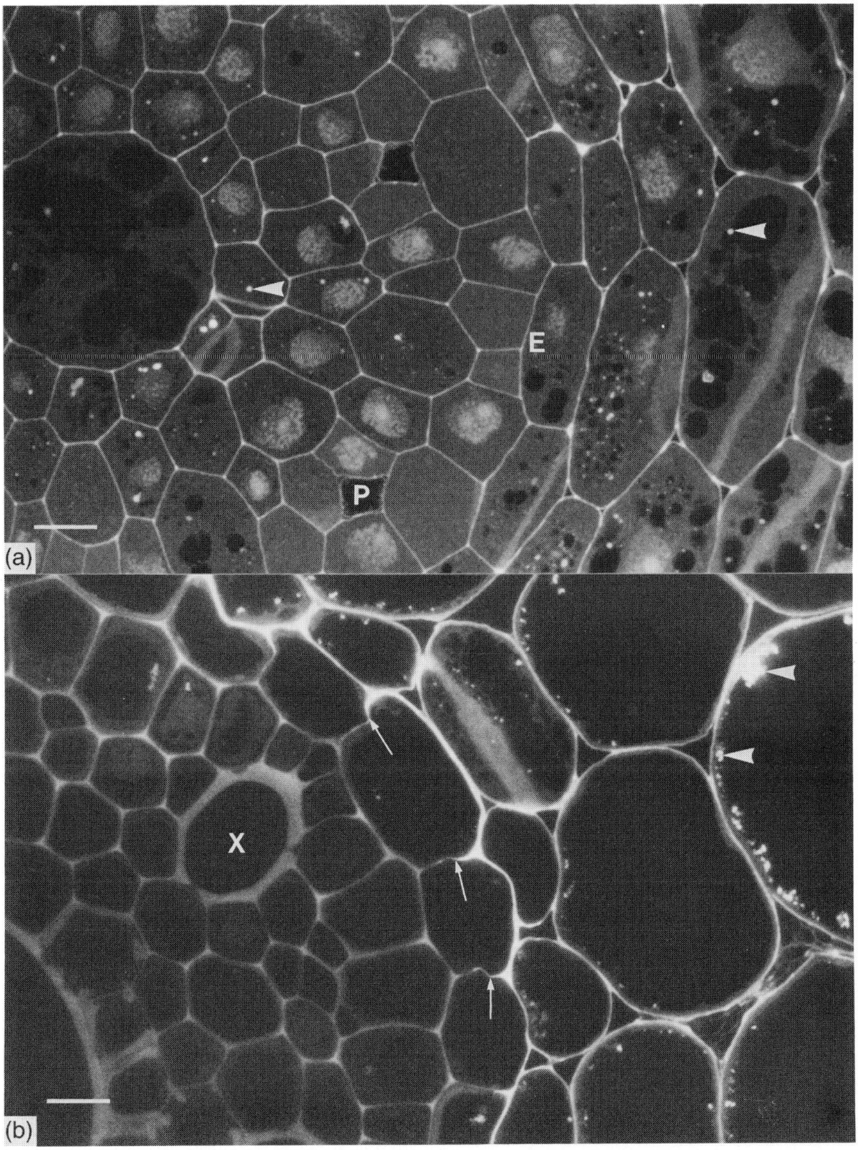

to a 1 cm zone approximately 6 cm behind the root tip. The dye has penetrated the cortex as far as the Casparian band (arrows) on the radial walls of the endodermis, but has not entered the vascular stele. Dye deposits are seen in the cell walls and also within the enlarged vacuoles (arrowheads) of the cortical cells. X, xylem vessel. The bar represents 10 μm. (From Oparka *et al.* 1988.)

the endodermal cells prevented the movement of dye into the stele. However, LY-CH continued to enter the vacuoles of cortical cells (Fig. 5 (b)). The inability of LY-CH to cross the plasma membrane at the point of the Casparian band and enter the stele via the symplast provided confirmation of its membrane-impermeant nature and further circumstantial evidence for endocytotic uptake.

More recently, Hillmer *et al.* (1991) allowed excised leaves of *Commelina communis* to transpire LY-CH. The dye entered the leaf by the xylem and was sequestered in the mesophyll cell vacuoles. When LY-CH was introduced into the cytoplasm of isolated mesophyll-cell protoplasts using a patch-clamp pipette they found that the dye could not enter the vacuole by an endocytotic route. Interestingly, LY-CH did not enter epidermal cells or guard cells but freely entered trichome cells, pointing to major differences in the ability to accumulate the dye.

Kinetics of LY-CH uptake into protoplasts

In order to characterise further the entry of LY-CH into plant cells Wright & Oparka (1989) examined the kinetics of LY-CH uptake into protoplasts from a number of plant species. Protoplasts of different species varied in their abilities to accumulate the dye but all sequestered it in the central vacuole. The time course of LY-CH uptake (Fig. 6) showed a rapid initial phase followed by a slower linear component. These kinetics are similar to those reported for animal (Swanson *et al.* 1985) and yeast (Riezman 1985) cells. The uptake of LY-CH was linear at concentrations of dye above 1 mg ml^{-1}, a feature consistent with fluid-phase endocytosis. Low temperature treatment (0 °C) removed the linear component of uptake but a small saturable component was retained (Fig. 7). This temperature-insensitive component has not as yet been identified and is evident also in yeast (Riezman 1985). The suggestion by Riezman that it represents dye retained within the wall is not tenable with protoplasts. The uptake of LY-CH into protoplasts was also sensitive to the inhibitors, 2-deoxy-D-glucose (a potent inhibitor of fluid-phase endocytosis in animal cells; Swanson *et al.* 1985) and sodium orthovanadate (a plasma membrane ATPase inhibitor). In addition, the protonophore carbonyl cyanide *m*-chlorophenylhydrazone (CCCP) did not significantly affect total uptake of the dye but resulted in a marked alteration in its intracellular compartmentation, causing the dye to be retained in numerous small vesicles rather than the central vacuole (Wright & Oparka 1989). Furthermore, calculations of the apparent membrane permeability of LY-CH gave values which were too low to be consistent with simple diffusion of the undissociated form of the molecule. Wright & Oparka (1989) also

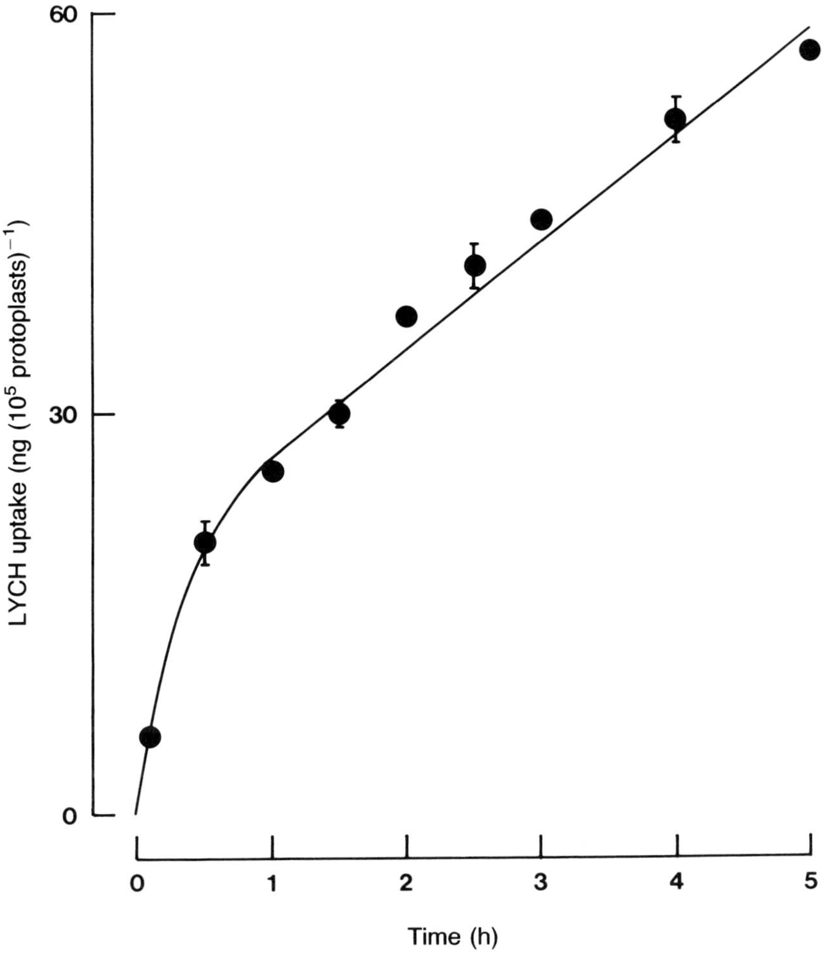

Fig. 6. The uptake of LY-CH into protoplasts of *Nicotiana clevelandii* with respect to time. Data are presented as means ± S.E.M. for three replicates. The regression for all data points was: uptake = 27.62 ± 16.17 ℓnt, where t is incubation time (r=0.98; p<0.001). (From Wright & Oparka 1989.)

considered the possibility of carrier-mediated transport of LY-CH (see recent reports that LY-CH is sequestered from the cytosol by organic ion transporters on the lysosomal membrane; Steinberg *et al.* 1987, 1988) but concluded that fluid-phase endocytosis was the most likely candidate for dye entry across the plasma membrane.

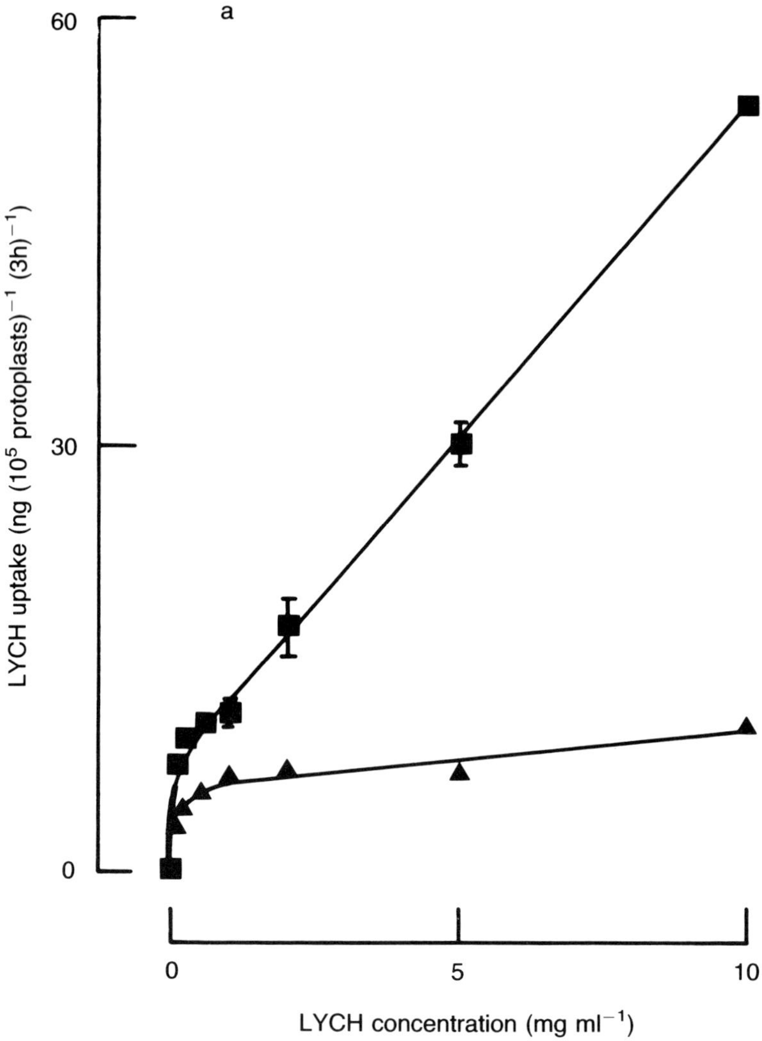

Fig. 7. The uptake of LY-CH with respect to substrate concentration at 25 °C (■) and 0 °C (▲). Data are presented as means ± S.E.M. for three replicates. (From Wright & Oparka 1989.)

Uptake of LY-CH into suspension-cultured cells

Fine suspension cultures of plant cells are convenient systems to use in fluid-phase endocytotic studies, since they have the practical advantage that all cells are simultaneously exposed to the same concentration of

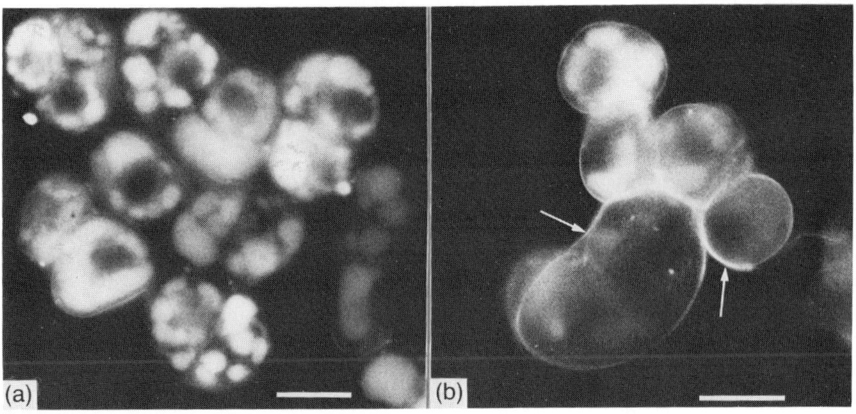

Fig. 8. Internalisation of LY-CH into suspension-cultured cells. (a) Suspension-cultured carrot cells incubated with the K^+ salt of LY-CH (1 mg ml^{-1}) for 20 h at 25 °C. Intense fluorescence is seen in the vacuoles and the dye is also trapped in some regions of the wall/plasma membrane interface (arrow). The bar represents 20 μm. (b) Suspension-cultured maize cells incubated with the Li^+ salt of LY-CH. After 1 h fluorescence is observed in vacuoles and also accumulates at the wall/plasma membrane interface (arrows). The bar represents 20 μm.

substrate. In a study of LY-CH and LY-VS uptake by cultured carrot cells, Cole (1989) observed strong fluorescence in the central vacuoles and in tubular compartments, but only after prolonged incubation (9–18 h). Both probes also appeared to accumulate along the cell wall/plasma membrane interface and no fluorescence was detected in nuclei or in the cytosolic compartment (Fig. 8 (a)). Similar results were obtained with isolated protoplasts. It is interesting to note that the rates of uptake by carrot cells and protoplasts are markedly slower than those reported for intact tissues and protoplasts from other species (Wright & Oparka 1989, see above). Hillmer *et al.* (1989) reported a similar time course for uptake and distribution of LY-CH in cultured carrot cells, but the probe did not accumulate in the central vacuole of protoplasts. Recently we have observed (L. Cole and coworkers, unpublished results) uptake of Lucifer dyes into cultured maize cells after shorter incubation periods of 2–5 h (Fig. 8 (b)). However, in another cell culture system (*Morinda citrifolia*) LY-CH apparently accumulates only in the cytosol (D. O'Driscoll, personal communication).

Another interesting observation (L. Cole and coworkers, unpublished results) is that the uptake of LY-CH by carrot cells and its sequestration in vacuoles is markedly reduced by the uricosuric drug Probenecid.

Although the site(s) of Probenecid action in these cells is(are) not known, Probenecid has been shown to block the uptake of LY-CH from the cytosol into lysosomes of macrophages probably by inhibiting an organic anion transporter (Steinberg *et al.* 1987, 1988). Goodwin *et al.* (1990) have shown the direct sequestration of microinjected LY-CH from the cytoplasm into vacuoles of leaves of *Egeria densa*, thus implying that plant cells are capable of transporting Lucifer Yellow from cytoplasm to vacuole. The mechanism of such vacuolar transport clearly deserves further study.

Other fluorescent probes for studying fluid-phase endocytosis

The synthesis of novel fluorescent probes with membrane-impermeant properties is an ever-expanding area of research, and several new probes have appeared in recent years. These include the newly-synthesised Cascade Blue hydrazide, an analogue of methoxypyrinetrisulphonic acid (Fig. 1 (c), see also Haugland 1989), which is a low molecular weight molecule that fluoresces exclusively under ultraviolet excitation. This probe can be used as a marker in conjunction with LY-CH for studies of fluid-phase endocytosis (see below).

As described above for FITC, many low molecular weight fluorochromes can be conjugated to macromolecules such as dextrans, proteins and peptides. These fluorochromes include the Lucifer dyes, Cascade Blue, rhodamine and the sulphonyl derivative of rhodamine, Texas Red (see Haugland 1989). Such probes have already found considerable use in studies of fluid-phase endocytosis in animal cells.

A wide range of fluorescent peptides are in common use in microinjection studies (see e.g. Goodwin 1983; Goodwin & Erwee 1985; Wolf *et al.* 1989) and are likely to find several potential applications in studies of fluid-phase endocytosis. For example FITC-Phe3 (M_r 850) is too large to pass through the plasmodesmata interconnecting mesophyll cells of *Nicotiana clevelandii* (P. Derrick, unpublished results). However, the same molecule will readily enter isolated mesophyll protoplasts. Extraction and chromatography of the internalised probe confirmed that the dye had entered in the conjugated form (K.M. Wright & P. Derrick, unpublished results).

Fluid-phase endocytosis of fluorescent probes during osmotic shock

Endocytosis has been implicated in the rapid recovery of plasma membrane during osmotic shock in higher plant cells (Withers & Cocking

1972). In isolated protoplasts, smooth endocytotic vesicles have been shown to form as a consequence of the membrane reduction associated with osmotic shock during freezing injury (Gordon-Kamm & Steponkus 1984*a,b*; see also Steponkus, this volume). Oparka *et al.* (1990) have shown recently that similar vesicles form during the rapid plasmolysis of intact onion epidermal cells and that such vesicles can be loaded with a succession of membrane-impermeant probes, provided that the probe is present in the apoplast at the onset of plasmolysis. Plasmolysis in the presence of LY-CH resulted in the uptake of this probe into numerous endocytotic vesicles. The dye-filled vesicles remained attached to the plasma membrane during deplasmolysis of the cells and no leakage was observed to the cytoplasm. Within minutes, however, several of the vesicles left their parietal position within the cell and underwent vigorous streaming in the cytoplasm. The movements of the vesicles were recorded by video-microscopy and their pathways followed within transvacuolar strands (Fig. 9 (a) to (c)). The continuous motion of cytoplasm towards and around the nucleus resulted in several of the larger (>0.5 μm) vesicles becoming trapped in the thin layer of cytoplasm surrounding the nucleus (Fig. 9 (d) and (e)).

Furthermore, a second plasmolysis/deplasmolysis (plasmolytic cycle) in the presence of Cascade Blue hydrazide resulted in the uptake of this probe into a second discrete population of fluid-phase vesicles. These also underwent cytoplasmic streaming (Fig. 9 (f) to (g)). Osmotically induced endocytotic loading would appear to be a potential strategy for the delivery of membrane-impermeant molecules to the cell interior, provided that a method can be found with which to discharge the vesicle contents to the cytosol.

Endocytotic vesicles, induced osmotically and observed over a period of several hours, did not discharge their contents into the central vacuole. This implies that a different mechanism of internalisation was in operation compared to that observed under isotonic conditions where probes are sequestered into the central vacuole (Wright & Oparka 1989; Hillmer *et al.* 1989). These observations point to major differences in the fate of endocytosed probes in the presence and absence of osmotic shock.

Conclusions

It is clear that plasmolysis of plant cells induced by hyperosmolarity causes endocytotic vesiculation which traps some of the ambient fluid during vesicle formation. These vesicles do not fuse with other cellular compartments, probably because they do not possess molecular signals which can be recognised by the membrane fusion machinery of the cell (see Hawes *et al.*, this volume). Although this differs from the typical

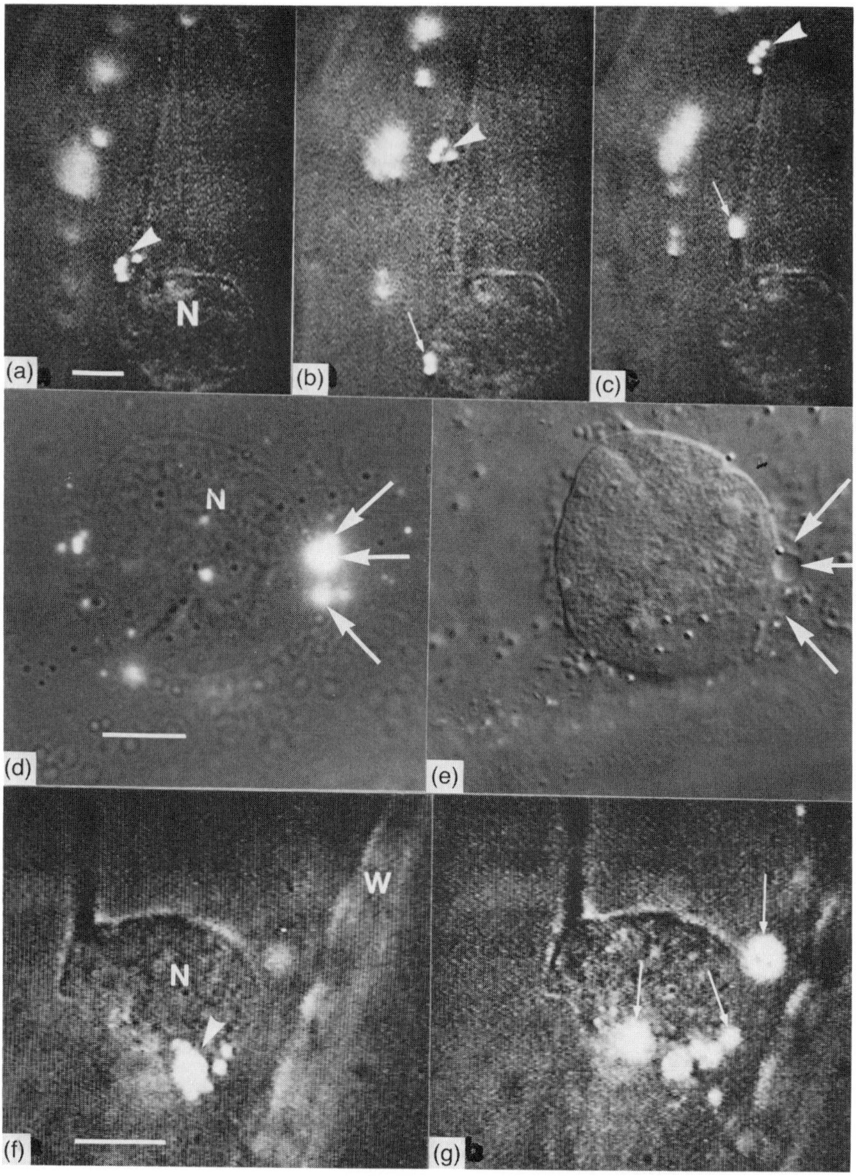

Fig. 9. Video-microscopy of osmotically induced endocytotic vesicles in onion epidermis. (a)–(c) Video sequence showing the movement of a group of endocytotic vesicles (arrowheads) within a transvacuolar strand in an intact onion epidermis cell. A second pair of vesicles moves past the nucleus in the same sequence (arrows). The frames were 7 s apart.

fluid-phase endocytotic pathway, it is still of potential significance particularly with respect to freezing injury (see Steponkus, this volume), wilting and exploitation as a means of delivery of selected molecules into plant cells.

From current experimental observations there is *prima facie* evidence for the occurrence of fluid-phase endocytosis in plant cells under normal osmotic conditions. The best evidence comes from the studies with LY-CH and is based on the assumption that the probe cannot enter the cytoplasm by direct penetration of the plasma membrane. However, observations that this fluorochrome partitions preferentially into the vacuolar system when introduced directly into the cytoplasm, cautions that the exclusive vacuolar localisation of the externally added probe is insufficient evidence that fluid-phase endocytosis is occurring. At present we cannot rule out the possibility that direct uptake of Lucifer Yellow occurs by a transport pathway other than, or in parallel with, fluid-phase endocytosis.

If fluid-phase endocytosis occurs in plant cells, it still remains to be tested whether clathrin is completely or partially involved in the process. If there is a non-clathrin-mediated pathway then there must be some other mechanism which drives vesiculation.

References

Baron-Epel, O., Gharyal, P.K. & Schindler, M. (1988). Pectins as mediators of wall porosity in soybean cells. *Planta*, **175**, 389–95.

Buckmaster, M.J., lo Braico, D., Ferris, A.L. & Storrie, B. (1987). Retention of pinocytozed solute by CHO cell lysosomes correlates with molecular weight. *Cell Biology International Reports*, **11**, 501–7.

Fig. 9 (*contd*)

N, nucleus. The bar represents 10 μm. (From Oparka *et al.* 1990.) (d) and (e) Endocytotic vesicles containing LY-CH (arrows) adjacent to the cell nucleus. (d) Combined fluorescence and bright field; (e) Differential interference contrast optics. Note that other subcellular particles show no fluorescence. The bar represents 10 μm. (From Oparka *et al.* 1990.) (f) and (g) Presence of two populations of endocytotic vesicles adjacent to the nucleus. The vesicles in (f) (arrowhead) contain LY-CH and were introduced during the first plasmolytic cycle (blue excitation). Those in (g) (arrows) contain Cascade Blue hydrazide and were introduced during the second plasmolytic cycle (ultraviolet excitation). Note that the LY-CH-containing vesicles remain faintly visible under ultraviolet excitation. W, cell wall. The bar represents 10 μm. (From Oparka *et al.* 1990.)

Chalmers, J. (1988). Plant plasma membrane activities. Ph.D. thesis, University of Oxford.

Cole, L. (1989). Endocytosis and the transport of fluorescent probes in suspension-cultured plant cells. M.Sc. thesis, University of Oxford.

Cole, L., Coleman, J., Evans, D. & Hawes, C. (1990). Internalisation of fluorescein isothiocyanate and fluorescein isothiocyanate-dextran by suspension-cultured plant cells. *Journal of Cell Science*, **96**, 721–30.

Coleman, J., Evans, D. and Hawes, C. (1988). Plant coated vesicles. *Plant Cell and Environment*, **11**, 669–84.

Daukas, G. & Zigmond, S.H. (1985). Inhibition of receptor-mediated but not fluid-phase endocytosis in polymorphonuclear leukocytes. *Journal of Cell Biology*, **101**, 1673–5.

Goodwin, P.B. (1983). Molecular size limit for movement in the symplast of the *Elodea* leaf. *Planta*, **157**, 124–30.

Goodwin, P.B. & Erwee, M.G. (1985). Intercellular transport studies by micro-injection methods. In *Botanical Microscopy*, ed. A.W. Robards, pp. 335–8. Oxford: Oxford University Press.

Goodwin, P.B., Shepherd, V. & Erwee, M.G. (1990). Compartmentation of fluorescent tracers injected into the epidermal cells of *Egeria densa* leaves. *Planta*, **181**, 129–36.

Gordon-Kamm, W.J. & Steponkus, P.L. (1984*a*). The behaviour of the plasma membrane following osmotic contraction of isolated protoplasts. *Protoplasma*, **123**, 161–73.

Gordon-Kamm, W.J. & Steponkus, P.L. (1984*b*). The influence of cold acclimation on the behaviour of the plasma membrane following osmotic contraction of isolated protoplasts. *Protoplasma*, **123**, 161–73.

Haugland, R. (1989). *Handbook of Fluorescent Probes and Research Chemicals*. Molecular Probes, Inc. Eugene, Or.

Heuser, J.E. & Anderson, R.G.W. (1989). Hypertonic media inhibit receptor-mediated endocytosis by blocking clathrin-coated pit formation. *Journal of Cell Biology*, **108**, 389–400.

Hillmer, S., Hedrich, R., Robert-Nicoud, M. & Robinson, D.G. (1991). Uptake of Lucifer Yellow CH in leaves of *Commelina communis* is mediated by endocytosis. *Protoplasma* (in press).

Hillmer, S., Quader, H., Robert-Nicoud, M. & Robinson, D.G. (1989). Lucifer Yellow uptake in cells and protoplasts of *Daucus carota* visualized by laser scanning microscopy. *Journal of Experimental Botany*, **40**, 417–23.

Horn, M.A., Heinstein, P.F. & Low, P.S. (1989). Receptor-mediated endocytosis in plant cells. *Plant Cell*, **1**, 1003–9.

Lang, T. & de Chastellier, C. (1985). Fluid phase and mannose receptor-mediated uptake of horseradish peroxidase in mouse bone marrow-derived macrophages. Biochemical and ultrastructural studies. *Biology of the Cell*, **53**, 149–54.

Larkin, J.M., Donzell, W.C. & Anderson, R.G.W. (1986). Potassium-dependent assembly of coated pits: new coated pits form as planar clathrin lattices. *Journal of Cell Biology*, **103**, 2619–27.

Madshus, I.H., Sandvig, K., Olsnes, S. & van Deurs, B. (1987). Effect of reduced endocytosis induced by hypertonic shock and potassium depletion on the infection of Hep 2 cells by picornaviruses. *Journal of Cellular Physiology*, **131**, 14–22.

Makarow, M. (1985). Endocytosis in *Saccharomyces cerevisiae*: internalization of α-amylase and fluorescent dextran into cells. *EMBO Journal*, **4**, 1861–6.

Makarow, M. & Nevalainen, L.T. (1987). Transport of a fluorescent macromolecule via endosomes to the vacuole in *Saccharomyces cerevisiae*. *Journal of Cell Biology*, **104**, 67–75.

Miller, D.K., Griffiths, E., Lenard, J. & Firestone, R.A. (1983). Cell killing by lysosomotropic detergents. *Journal of Cell Biology*, **97**, 1841–51.

Oka, J.A. & Weigel, P.H. (1987). Hyperosmolarity preferentially inhibits galactosyl receptor mediated but not fluid-phase endocytosis in isolated rat hepatocytes. *Journal of Cell Biology*, **105**, 1756.

Oparka, K.J. & Prior, D.A.M. (1988). Movement of Lucifer Yellow CH in potato tuber storage tissues: a comparison of symplastic and apoplastic transport. *Planta*, **176**, 533–40.

Oparka, K.J., Prior, D.A.M. & Harris, N. (1990). Osmotic induction of fluid-phase endocytosis in onion epidermal cells. *Planta*, **180**, 555–61.

Oparka, K.J., Robinson, D., Prior, D.A.M., Derrick, P. & Wright, K.M. (1988). Uptake of Lucifer Yellow CH into intact barley roots: evidence for fluid-phase endocytosis. *Planta*, **176**, 541–7.

Preston, R.A., Murphy, R.F. & Jones, E.W. (1987). Apparent endocytosis of fluorescein isothiocyanate-conjugated dextran by *Saccharomyces cerevisiae* reflects uptake of low molecular weight impurities not dextran. *Journal of Cell Biology*, **105**, 1981–7.

Riezman, H. (1985). Endocytosis in yeast: several of the yeast secretory mutants are defective in endocytosis. *Cell*, **40**, 1001–9.

Riezman, H., Chvatchko, Y. & Dulic, V. (1986). Endocytosis in yeast. *Trends in Biochemical Sciences*, **11**, 325–8.

Robinson, D.G. & Hillmer, S. (1990). Endocytosis in plants. *Physiologica Plantarum*, **79**, 96–104.

Sandvig, K., Olsnes, S., Brown, J.E., Petersen, O.W. & van Deurs, B. (1989). Endocytosis from coated pits of Shiga toxin: a glycolipid-binding protein from *Shigella dysenteriae*. I. *Journal of Cell Biology*, **108**, 1331–44.

Sandvig, K., Olsnes, S., Petersen, O.W. & van Deurs, B. (1987). Acidification of the cytosol inhibits endocytosis from coated pits. *Journal of Cell Biology*, **105**, 679–89.

Steinberg, T.H., Newman, A.S., Swanson, J.A. & Silverstein, S.C.

(1987). Macrophages possess probenecid-inhibitable organic anion transporters that remove fluorescent dyes from the cytoplasmic matrix. *Journal of Cell Biology*, **105**, 2695–702.

Steinberg, T.H., Swanson, J.A. & Silverstein, S.C. (1988). A prelysosomal compartment sequesters membrane-impermeant fluorescent dyes from the cytoplasmic matrix of J774 macrophages. *Journal of Cell Biology*, **107**, 887–96.

Steinman, R.M., Mellman, I.S., Muller, W.A. & Cohn, Z.A. (1983). Endocytosis and recycling of the plasma membrane. *Journal of Cell Biology*, **96**, 1–27.

Stewart, W.W. (1978). Functional connections between cells as revealed by dye-coupling with a highly fluorescent naphthalimide tracer. *Cell*, **14**, 741–59.

Stewart, W.W. (1981). Lucifer dyes – highly fluorescent dyes for biological tracing. *Nature* (*London*), **292**, 17–21.

Swanson, J. (1989). Fluorescent labeling of endocytic compartments. In *Methods in Cell Biology*, ed. Y.L. Wang & D. Lansing Taylor, pp. 137–51. New York: Academic Press Inc.

Swanson, J., Burke, E. & Silverstein, S.C. (1987). Tubular lysosomes accompany stimulated pinocytosis in macrophages. *Journal of Cell Biology*, **104**, 1217–22.

Swanson, J.A., Yirinec, B.D. & Silverstein, S.C. (1985). Phorbol esters and horseradish peroxidase stimulate pinocytosis and redirect the flow of pinocytosed fluid in macrophages. *Journal of Cell Biology*, **100**, 851–9.

van Deurs, B., Röphe, C. & Thorball, N. (1984). Kinetics of pinocytosis studied by flow cytometry. *European Journal of Cell Biology*, **34**, 96–102.

van Deurs, B., Petersen, O.W., Olsnes, S.B., Sandvig, K. (1989). The ways of endocytosis. *International Review of Cytology*, **117**, 131–77.

Withers, L.A.W. & Cocking, E.C. (1972). Fine-structural studies on spontaneous and induced fusion of higher plant protoplasts. *Journal of Cell Science*, **11**, 59–75.

Wolf, S., Deom, C.M., Beachy, R.N. & Lucas, W.J. (1989). Movement of protein of tobacco mosaic virus modifies plasmodesmatal size exclusion limit. *Science*, **246**, 377–9.

Wright, K.M. & Oparka, K.J. (1989). Uptake of Lucifer Yellow CH into plant cell protoplasts: a quantitative assessment of fluid-phase endocytosis. *Planta*, **179**, 257–64.

P.L. STEPONKUS

Behaviour of the plasma membrane during osmotic excursions

Under favourable environmental conditions, plant cells function under conditions of positive turgor and are not subjected to large osmotic excursions. However, under conditions of environmental stress such as drought or freezing, cell dehydration occurs and large osmotic excursions are possible, with the magnitude of freeze/thaw-induced osmotic excursions being the largest. For example, in many crop species, drought stress is usually manifested over a range of water potentials from −1.0 to −2.0 MPa, with a loss of turgor occurring when approximately 10% to 15% of the osmotically active water is removed from the cells (Hsiao 1973). In contrast, when seedlings of winter cereals are frozen to temperatures over the range of −2 to −20 °C, the cells are subjected to water potentials ranging from −2.4 to −24 MPa and more than 90% of the osmotically active water is removed from the cells (Steponkus 1980).

Cell dehydration occurs during freezing because ice formation occurs initially in the extracellular matrix and creates a gradient between the chemical potential of the intracellular and extracellular solutions, with the chemical potential of ice being lower than that of water at the same temperature. Water efflux will continue until the intracellular solutes are sufficiently concentrated that the chemical potential of the intracellular solution is equal to that of the extracellular ice. The probability of intracellular ice formation is low at temperatures between 0 °C and −10 °C because, in this temperature range, intracellular ice-nucleating agents are not effective and the plasma membrane serves as a barrier to the extracellular ice crystals to prevent seeding of the supercooled intracellular solution. Because cell volume decreases linearly as a function of the reciprocal of the chemical potential (osmolality), large volumetric reductions occur at relatively high subzero temperatures. For example, following thermal and osmotic equilibration at −5 °C, which is equivalent to an osmotic potential of −6 MPa, more than 80% of the osmotically active water is removed from the cells. During thawing, the process is reversed and the cells will regain their initial volume if injury has not occurred.

Because the plasma membrane is a primary site of freezing injury (see Steponkus 1984; Steponkus & Lynch 1989*a*), the behaviour of the plasma membrane during large osmotic excursions and the effect of cold acclimation on this behaviour are of major interest. To study the intrinsic properties of the plasma membrane, we have elected to study the behaviour of isolated protoplasts, especially mesophyll protoplasts isolated from leaves of winter rye (*Secale cereale* L. cv. Puma). This report is a synopsis of studies in my laboratory of the behaviour of the plasma membrane during osmotic excursions and is presented in a manner that reflects the evolution of our understanding – from the phenomenology to the molecular mechanisms.

Osmotic behaviour of isolated protoplasts

Isolated protoplasts behave as ideal osmometers and exhibit characteristic Boyle–van't Hoff behaviour over a wide range of osmolalities, i.e. their volume varies linearly as a function of osmolality^{-1} (Fig. 1). For protoplasts isolated from non-acclimated rye leaves, the fractional volume (FV) expressed as the Boyle–van't Hoff relationship is FV = 0.084 + 0.485 OSM^{-1} as measured over the range of 0.3 to 3.0 osmolal (Dowgert & Steponkus 1983). For protoplasts isolated from cold-acclimated rye leaves, FV = 0.146 + 0.879 OSM^{-1} over the range 0.4 to 6.0 osmolal, with the increased slope a consequence of the increase in the internal solute concentration. A 0.53 osmolal solution is isotonic for protoplasts isolated from non-acclimated rye leaves, and 1.03 osmolal is isotonic for protoplasts isolated from cold-acclimated leaves.

During osmotic expansion in hypotonic solutions, lysis of isolated protoplasts is stochastic. In a suspension of protoplasts isolated from non-acclimated rye leaves, lysis occurs in 50% of the protoplasts after transfer to a 0.34 osmolal solution, which results in an increase in the fractional volume to 1.56 (Dowgert & Steponkus 1984). Although osmometric behaviour is usually expressed as a function of volume, lysis of the plasma membrane occurs when a critical surface area is exceeded. For example, during an isotonic–hypotonic osmotic excursion of rye protoplasts, 50% of the protoplasts lyse when a fractional surface area of 1.34 is exceeded. However, the critical surface area is not a constant: Tornava (1939) and Scarth *et al.* (1940) observed that following osmotic contraction in hypertonic solutions, protoplasts lysed upon return to an isotonic solution.

Quantitative studies of this phenomenon were first reported for mesophyll protoplasts of spinach (Wiest & Steponkus 1978) and subsequently for rye protoplasts (Dowgert & Steponkus 1984). For both, the

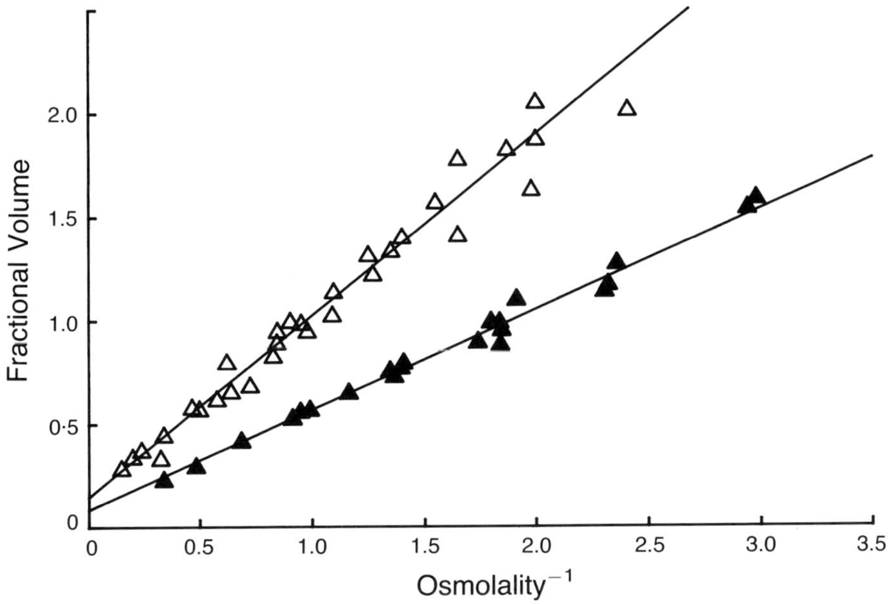

Fig. 1. Boyle–van't Hoff plot of the fractional volume of rye protoplasts as a function of osmolality^{-1}. The protoplasts were isolated from either non-acclimated or cold-acclimated rye leaves. Diameters of approximately 100 protoplasts were measured at 22 different osmolalities over the range 0.3 to 3.0 osmolal for protoplasts isolated from non-acclimated leaves (▲) and at 31 different osmolalities over the range 0.4 to 6.0 osmolal for protoplasts isolated from cold-acclimated leaves (△). (From Dowgert & Steponkus 1983.)

critical surface area at which lysis occurs depends on the previous osmotic history of the protoplasts: the surface area at which lysis occurs decreases if the protoplasts have previously been subjected to osmotic contraction, and sufficiently large volumetric contractions are irreversible. For rye protoplasts, lysis occurs at a fractional surface area of 1.34 following transfer from an isotonic (0.53 osmolal) to a hypotonic solution (Fig. 2 (a)). However, if the protoplasts are first contracted in a hypertonic (1.02 osmolal) solution and then transferred to a more dilute solution, lysis occurs at a fractional surface area of 1.04. If the protoplasts are first contracted in a 1.20 osmolal solution, lysis occurs before the protoplasts regain their isotonic surface area, i.e. at a fractional surface area of 0.86. Similar values have also been obtained with spinach protoplasts. Thus, neither rye nor spinach protoplasts lyse at a constant critical surface area. Rather, lysis occurs at a constant surface area increment (i.e. the

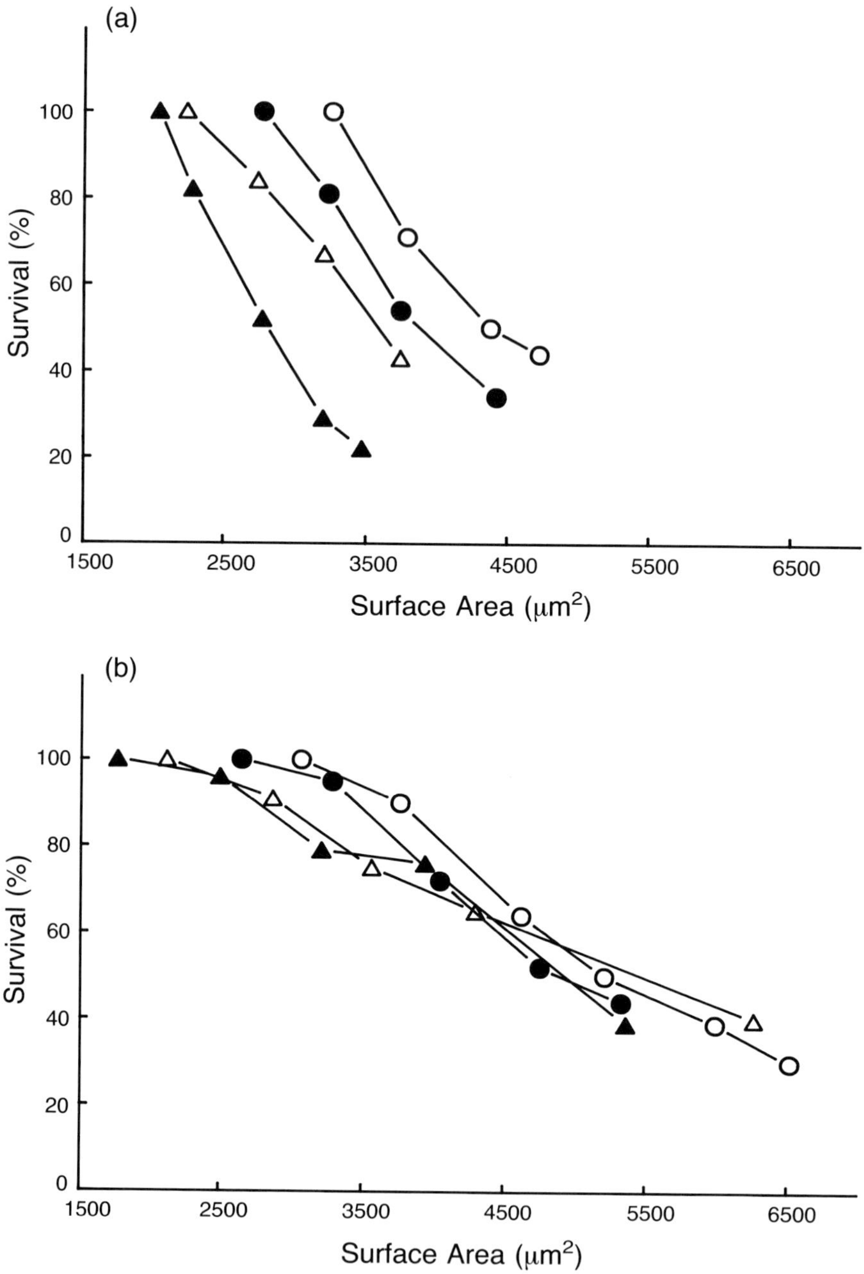

Fig. 2. Survival of protoplasts as a function of the change in the surface area. (a) Protoplasts isolated from non-acclimated rye leaves: expansion

difference between the area at lysis and the area in the hypertonic solution), which is independent of the extent of osmotic contraction. This value has been termed the tolerable surface area increment (TSAI, Steponkus & Wiest 1979), with the $TSAI_{50}$ the average tolerable surface area increment at which lysis occurs in 50% of the protoplasts. This value is 0.34 (in fractional units) for both spinach and rye protoplasts (cf. Wiest & Steponkus 1978; Dowgert & Steponkus 1984) and is independent of the protoplast size. Thus, both spinach and rye protoplasts will tolerate a 34% increase in the surface area of the plasma membrane before lysis occurs – regardless of the initial size in an isotonic solution or if the size is reduced by osmotic contraction.

Following cold acclimation, the behaviour of the plasma membrane is substantially altered (Steponkus *et al.* 1981). For protoplasts isolated from cold-acclimated rye leaves, 50% of the protoplasts survive transfer from an isotonic solution (1.03 osmolal) to a 0.42 osmolal solution, which results in a fractional volume of 2.22 and a fractional surface area of 1.70 (Fig. 2 (b)). More important, osmotic contraction of protoplasts isolated from cold-acclimated leaves is readily reversible, and the relationship of survival as a function of the global surface area is similar to that of protoplasts expanded from isotonic solutions (Dowgert & Steponkus 1984).

These observations prompt several questions: what determines the limits of expansion from isotonic solutions? Why is there a nearly two-fold difference in the surface area expansion potential of protoplasts isolated from non-acclimated rye leaves (0.34) and that of protoplasts isolated from cold-acclimated leaves (0.70)? Why are sufficiently large osmotic contractions irreversible in protoplasts isolated from non-acclimated leaves, but reversible in protoplasts isolated from cold-acclimated leaves? The answers to these questions have required an analysis of the mechanical properties of the plasma membrane and characterisation of its stress–strain relation, ultrastructural studies of the plasma membrane during osmotic excursions, and membrane engineering studies to establish structure–function relationships.

Fig. 2 (*contd*)

from an isotonic (0.53 osmolal) solution (○) or hypertonic solutions: 0.7 (●), 1.016 (△), or 1.20 (▲) osmolal. (b) Protoplasts isolated from cold-acclimated rye leaves: expansion from an isotonic (1.03 osmolal) solution (○) or hypertonic solutions: 1.33 (●), 2.04 (△), or 3.0 (▲) osmolal. (From Dowgert & Steponkus 1984.)

Mechanical properties of the plasma membrane

Protoplasts are spherical and have a turgid appearance over a wide range of volumes – characteristics which indicate that the plasma membrane is under tension and is supporting a hydrostatic pressure in both hypotonic and hypertonic solutions. Although it is obvious that the tension would increase during osmotic expansion, it is not obvious why the plasma membrane would be under tension after osmotic contraction. Also, although the limited expansion potential of red blood cells has been ascribed to viscoelastic behaviour (Evans & Skalak 1979), the large (± 30%) changes in surface area of protoplasts would not appear to be accommodated by only elastic expansion or contraction of the plasma membrane, which would require proportional changes in the thickness of the plasma membrane. Similarly, the observation that the expansion potential of isolated protoplasts is characterised by a constant surface area increment rather than a constant maximum surface area, suggests that the plasma membrane does not behave as a simple elastic body.

To resolve these questions, we characterised the mechanical properties and the stress–strain relation of the plasma membrane of isolated rye protoplasts (Wolfe & Steponkus 1981, 1983) by using micropipette aspiration as described by Mitchison & Swann (1954). In isotonic solutions, the plasma membrane of isolated protoplasts is under a relatively small 'resting tension' (γ_r) of 0.1 mN m^{-1} and sustains a hydrostatic pressure of approximately 1 kPa. During area changes over periods of less than 1 s, the plasma membrane behaves elastically and changes in the surface tension of the plasma membrane are related to changes in area by the equation $\gamma = k_A(\Delta A/A)$, where k_A is the area elastic modulus and $\Delta A/A$ is the fractional change in surface area. For rye protoplasts, the k_A of the plasma membrane is 230 mN m^{-1}. Given that the probability of lysis increases greatly at tensions of 4 to 6 mN m^{-1}, it is estimated that intrinsic elastic expansion of the plasma membrane is limited to 2% to 3%. Therefore, in order to accommodate the much larger increases in surface area that occur during osmotic expansion (i.e. 34% for protoplasts isolated from non-acclimated rye leaves and 70% for protoplasts isolated from cold-acclimated leaves), additional membrane material must be incorporated into the plane of the plasma membrane. Thus, in the first second of an osmotic excursion, small changes in surface area are predominantly elastic and result in large changes in γ because of the large value of k_A relative to γ_r. However, over time scales greater than 1 s, changes in γ control the movement of material into and out of the plasma membrane: if $\gamma > \gamma_r$, material is transferred into the plasma membrane; if $\gamma = 0$, material is deleted from the plasma membrane.

The dynamics of incorporation of membrane material into the plasma membrane and the probable frequency of lysis as a function of tension were studied initially by measuring changes in area during a linear increase in tension, which was effected by micropipette aspiration (Wolfe *et al.* 1985). The rate of the proportional increase in area (Z) increased exponentially with tension and was independent of the rate of increase in tension, which ranged from 138 to 2477 Pa min^{-1}. At tensions over the range 0.1 to 3.0 mN m^{-1}, Z was less than 0.05% s^{-1}; over the range 3.0 to 6.0 mN m^{-1}, Z increased from 0.05 to 2.0% s^{-1}. Typically, lysis occurred when the tension in the plasma membrane was in the range of 4 to 6 mN m^{-1}. However, lysis is not simply the result of the tension exceeding some critical value. Rather, it is a stochastic phenomenon, the probability of which increases exponentially with tension: the predicted lifetime is 43 s at γ=5.0 mN m^{-1} and is halved by each increase of 0.5 mN m^{-1}.

In studies in which a linear increase in tension is effected by micropipette aspiration, lysis is inevitable. In contrast, during osmotic expansion following transfer to a hypotonic solution, tension in the plasma membrane quickly reaches a maximum, but then declines as additional material is incorporated into the membrane (see Wolfe *et al.* 1986*a*). To obtain direct information of the stress–strain relation of the plasma membrane during osmotic expansion, the stress (tension) and the strain (change in area) were measured simultaneously during osmotic expansion (Wolfe *et al.* 1986*a*). For these studies, the protoplasts were subjected to a step-like change in tonicity of the suspending medium from isotonic to solutions of various tonicities, including transfer to distilled water. Somewhat unexpectedly, under these conditions, the range of areas over which lysis occurred was similar in all treatments – regardless of the osmolality of the suspending medium and the rate of expansion. Thus, lysis occurred after the same surface area increment, whether the protoplasts were transferred to a hypotonic solution (0.29, 0.24, 0.18, 0.14 osmolal) or distilled water. Intuitively, one would have expected that the protoplasts transferred to distilled water would have lysed at a smaller area. These results can be explained by the compensation between tension-dependent incorporation of membrane material and by lysis being a stochastic, tension-dependent phenomenon. At more rapid rates of expansion, the increase in tension is greater, but so is the rate at which material is incorporated into the membrane. At slower rates of expansion, the probability of lysis is less, but the duration of time over which lysis can occur is longer because the rate of incorporation is slower.

These studies were subsequently extended to consider the mechanics of injury in protoplasts that were first subjected to osmotic contraction before osmotic expansion (Dowgert *et al.* 1987). Under these conditions there is a substantial difference in the behaviour of protoplasts isolated from non-acclimated rye leaves in comparison to those isolated from cold-acclimated leaves. Upon exposure of protoplasts isolated from non-acclimated leaves to hypertonic solutions, γ_r decreased to zero with the initial efflux of water. At zero tension, membrane material was deleted from the flaccid plasma membrane in the form of endocytotic vesicles (see below). With the progressive decrease in membrane area over the range 1.0 to 0.55, the tension was re-established to the same value (0.1 mN m^{-1}) measured in isotonic solutions. In contrast, the resting tension of protoplasts isolated from cold-acclimated leaves decreased from 0.17 mN m^{-1} to zero and remained at zero following osmotic contraction. In this case, membrane material was not deleted from the plasma membrane. Rather, the surface area of the plasma membrane was conserved because of the formation of exocytotic extrusions (see below). As a result of this differential behaviour during contraction, there was a difference in the stress–strain relation during subsequent osmotic expansion. For protoplasts isolated from non-acclimated leaves, the tension in the plasma membrane increased rapidly after small increments in area. In contrast, tension in the plasma membrane of protoplasts isolated from cold-acclimated leaves did not increase until after the isotonic area was exceeded.

These studies established a causal relationship between (1) endocytotic vesiculation of the plasma membrane and the irreversibility of sufficiently large contractions in protoplasts isolated from non-acclimated leaves and (2) the formation of exocytotic extrusions and the reversibility of osmotic contraction in protoplasts isolated from cold-acclimated leaves. (For a more detailed discussion of the mechanics of osmotic contraction and expansion, see Wolfe *et al.* 1986; for a detailed analysis of the quantitative differences between protoplasts isolated from non-acclimated and cold-acclimated rye leaves, see Wolfe *et al.* 1985, 1986*a*; Dowgert *et al.* 1987.)

Endocytotic vesiculation of the plasma membrane during osmotic contraction

Osmotic contraction of isolated protoplasts results in endocytotic vesiculation of the plasma membrane. Initially, the process of endocytotic vesiculation was observed in spinach (Steponkus & Wiest 1979) and rye (Steponkus *et al.* 1982) protoplasts by using fluorescein-concanavalin A to label the plasma membrane prior to osmotic contraction. However, the process is best observed by using differential interference contrast optics and high-resolution video microscopy (Steponkus *et al.* 1983). Following an increase in the tonicity of the suspending medium and the efflux of

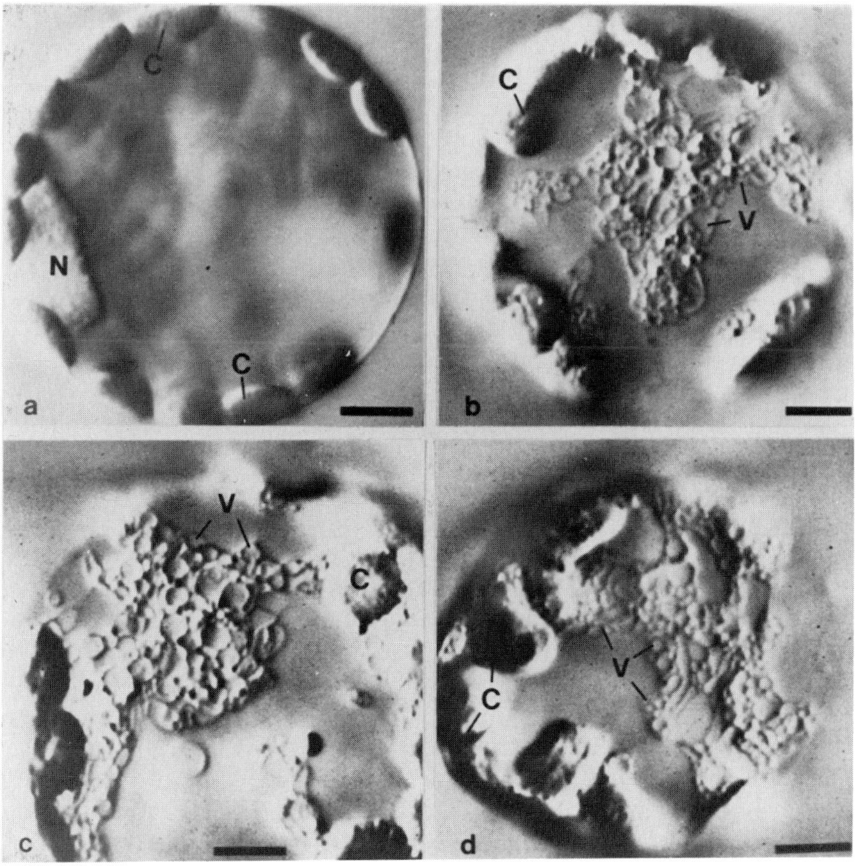

Fig. 3. Computer-enhanced video-microscopy of protoplasts isolated from non-acclimated rye leaves and suspended in either (a) isotonic (0.53 osmolal) or (b), (c) and (d) hypertonic (1.20 osmolal) sorbitol solutions. V, vesicles derived from the plasma membrane; C, chloroplast; N, nucleus; the bars represent 5 μm. (From Gordon-Kamm & Steponkus 1984*a*.)

water from the protoplast, the protoplast becomes flaccid and irregularly shaped. The plasma membrane begins to flutter and undulate; invagination and vesiculation of the plasma membrane occurs in time scales of seconds. The vesicles range in diameter from 0.3 to 1.0 μm and tend to occur in clusters subtending the plasma membrane (Fig. 3). This process continues for several minutes after osmotic equilibration has occurred. As material is removed from the plasma membrane, the protoplast returns to its spherical shape and again appears turgid. This process may occur over periods of time ranging from tens of seconds to tens of minutes, depending on the extent of osmotic contraction. Long ago, protoplasmatologists referred to the 'sphering-time' of protoplasts, which was the time required for a protoplast to regain its spherical shape following osmotic contraction. This phenomenon is a consequence of endocytotic vesiculation and the reduction in area of the plasma membrane.

In electron micrographs of rye protoplasts that are equilibrated in a hypertonic solution, the surface of the plasma membrane appears smooth (Fig. 4). In thin sections, the endocytotic vesicles appear in large clusters either subtending the plasma membrane or between chloroplasts that are adjacent to the plasma membrane (Fig. 5). Not all of the vesicles remain near the plasma membrane; some are observed throughout the cytoplasm (Gordon-Kamm & Steponkus 1984*a*). Following equilibration and the return to sphericity in the hypertonic solution, the plasma membrane appears smooth and free from any folds or pleats. The thickness of the plasma membrane is unchanged. In freeze-fracture studies, the morphology of the plasma membrane fracture faces remained unchanged following equilibration in a hypertonic (1.0 osmolal) solution that resulted in a 40% reduction in the surface area of the plasma membrane. The size and frequency of the intramembrane particles on both fracture planes was similar in both isotonic and hypertonic solutions. We interpret this to indicate that endocytotic vesiculation involves a unit-membrane deletion rather than a preferential deletion of either lipids or proteins.

As previously stated, sufficiently large area reductions are irreversible, i.e. protoplasts that are contracted in a sufficiently hypertonic solution will lyse upon return to an isotonic solution. This suggests that the endocytotic vesicles formed during osmotic contraction are not readily reincorporated back into the plasma membrane during subsequent osmotic expansion. Both video microscopy and electron microscopy studies of osmotic expansion are consistent with this interpretation.

Fig. 4. Scanning (a), (c), (e) and transmission (b), (d), (f) electron micrographs of protoplasts isolated from non-acclimated rye leaves. All bars represent 5 μm. (a) and (b) Protoplasts suspended in isotonic (0.53 osmolal) sorbitol; Va, vacuole; C, chloroplast; M, mitochondria. (c) and (d) Protoplasts suspended in hypertonic (1.20 osmolal) sorbitol. The surface (c) is similar to that observed in isotonic solutions (a). In thin sections (d), numerous vesicles (V) derived from the plasma membrane are observed in the cytoplasm. (e) and (f) Protoplasts equilibrated in a hypertonic sorbitol solution (1.20 osmolal) and subsequently suspended in a more dilute (0.60 osmolal) sorbitol solution. The surface texture (e)

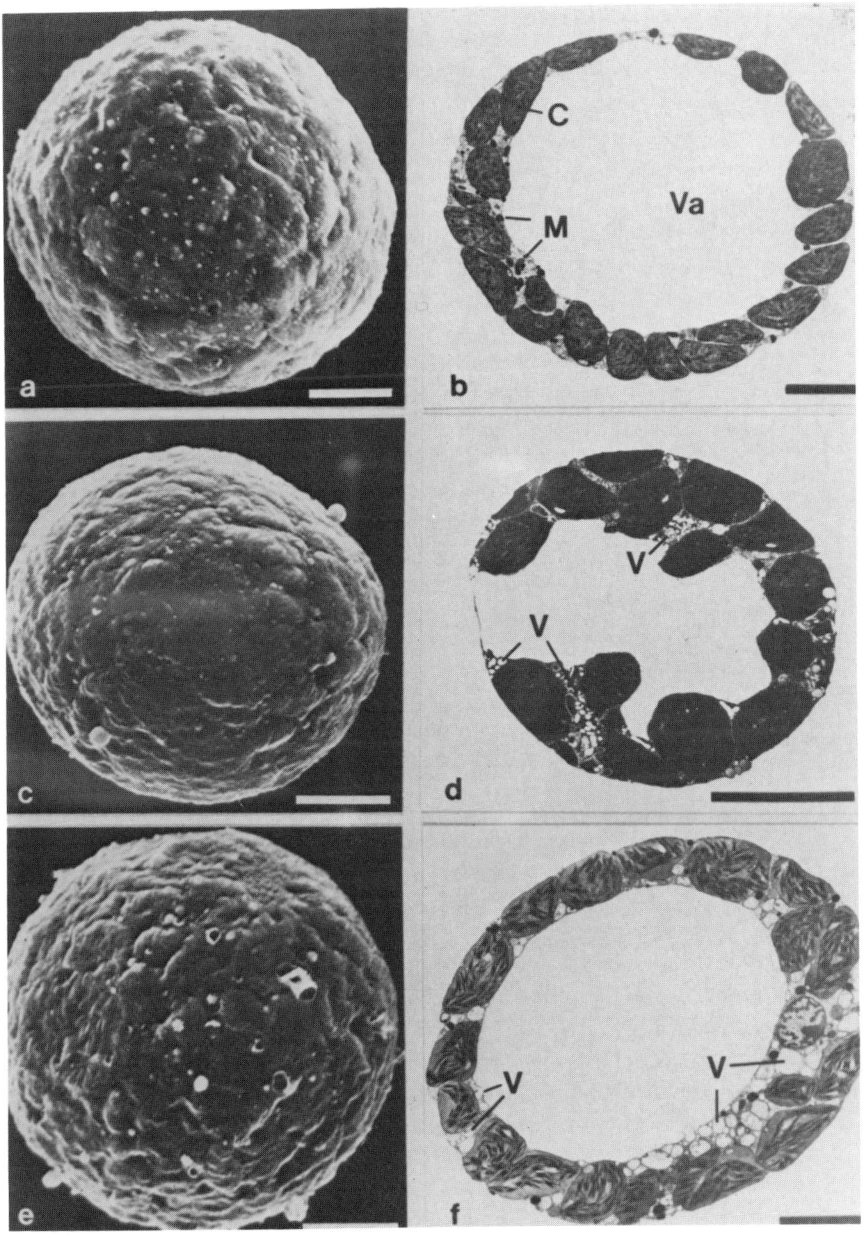

remains smooth. However, in thin sections (f) the plasma membrane vesicles are still present in the cytoplasm. (From Gordon-Kamm & Steponkus 1984*a*.)

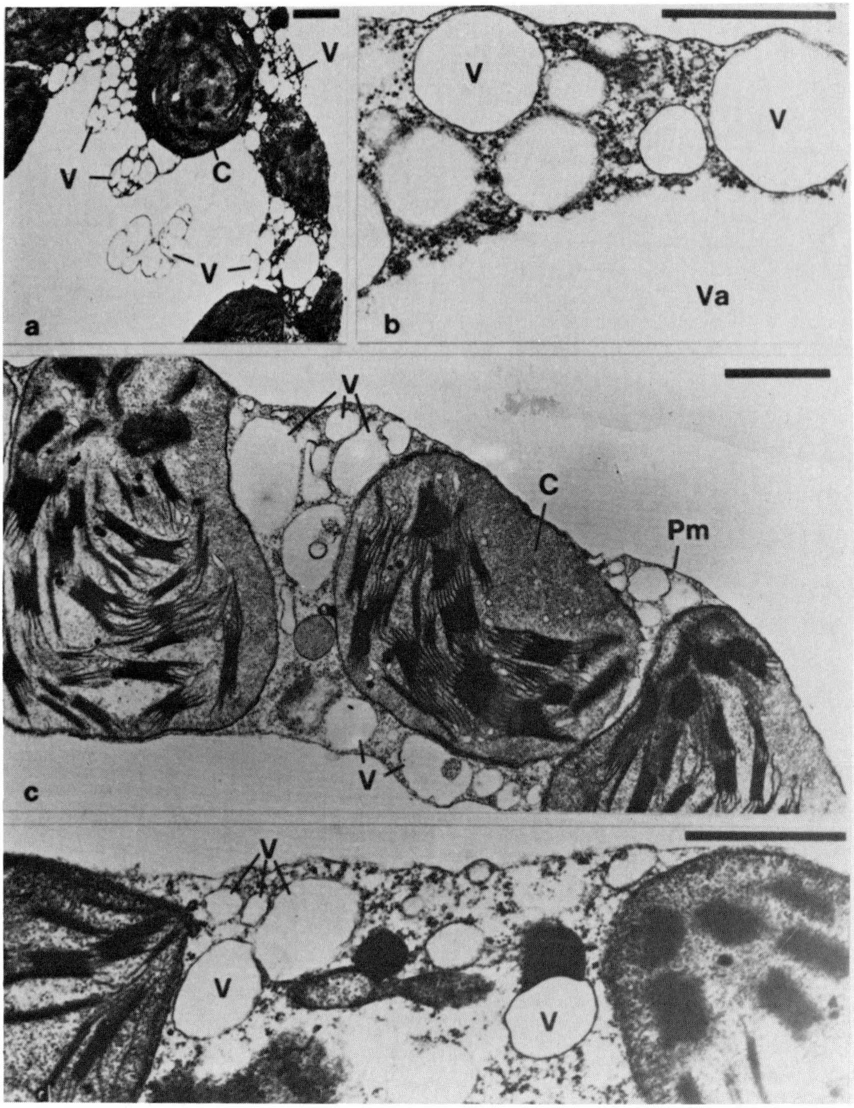

Fig. 5. Thin sections of protoplasts isolated from non-acclimated rye leaves and suspended in a hypertonic (1.20 osmolal) sorbitol solution. Vesicles (V) derived from the plasma membrane often appear in clusters (a), either subtending the plasma membrane (a and b) or between the chloroplasts (C) (c and d). (From Gordon-Kamm & Steponkus 1984*a*.)

When lysis occurs in protoplasts that have been transferred from a hypertonic solution to an isotonic solution, vesicles are discharged into the suspending medium following rupture of the plasma membrane. In electron micrographs of partially re-expanded protoplasts, endocytotic vesicles are still present in the cytoplasm. These observations and the fact that the expansion potential of the protoplasts is determined by a constant surface area increment regardless of the extent of contraction suggest that the endocytotic vesicles do not enter the membrane reservoir that is responsible for expansion.

Formation of exocytotic extrusions

Osmotic contraction of protoplasts isolated from cold-acclimated rye leaves results in the formation of exocytotic extrusions of the plasma membrane (Dowgert & Steponkus 1984; Gordon-Kamm & Steponkus 1984*b*). In scanning electron micrographs, the extrusions appear as tethered spheres or polyps on the surface of the protoplasts (Fig. 6). The diameter of the spheres is approximately 0.5 μm. In some instances, the tethers are very short and the spheres appear sessile; in other instances, the length of the tethers varies greatly, with some 5 μm in length (Fig. 7). In thin sections, the interior of the extrusions appears densely osmiophilic and has an amorphous, aparticulate appearance when cross-fractured (Fig. 8). The extrusions are bounded by the plasma membrane. Formation of the extrusions is readily reversible upon return to isotonic conditions, and there is no evidence of the osmiophilic granules.

Alterations in the freeze-fracture morphology of the plasma membrane occur also following osmotic contraction and the formation of exocytotic extrusions. The intramembrane particle density on the protoplasmic fracture face is significantly increased following osmotic contraction, but remains unchanged on the cytoplasmic fracture face. In addition, large areas of paracrystalline arrays appear on the protoplasmic fracture face. The increase in the intramembrane particle density on the protoplasmic fracture face was observed both in the regions of the plasma membrane that extended over the extrusions and in the regions between the extrusions. We have interpreted the increase in the intramembrane particle frequency as a result of the preferential subduction of lipid material from the plasma membrane and is presumed to be the densely osmiophilic material in the interior of the extrusions.

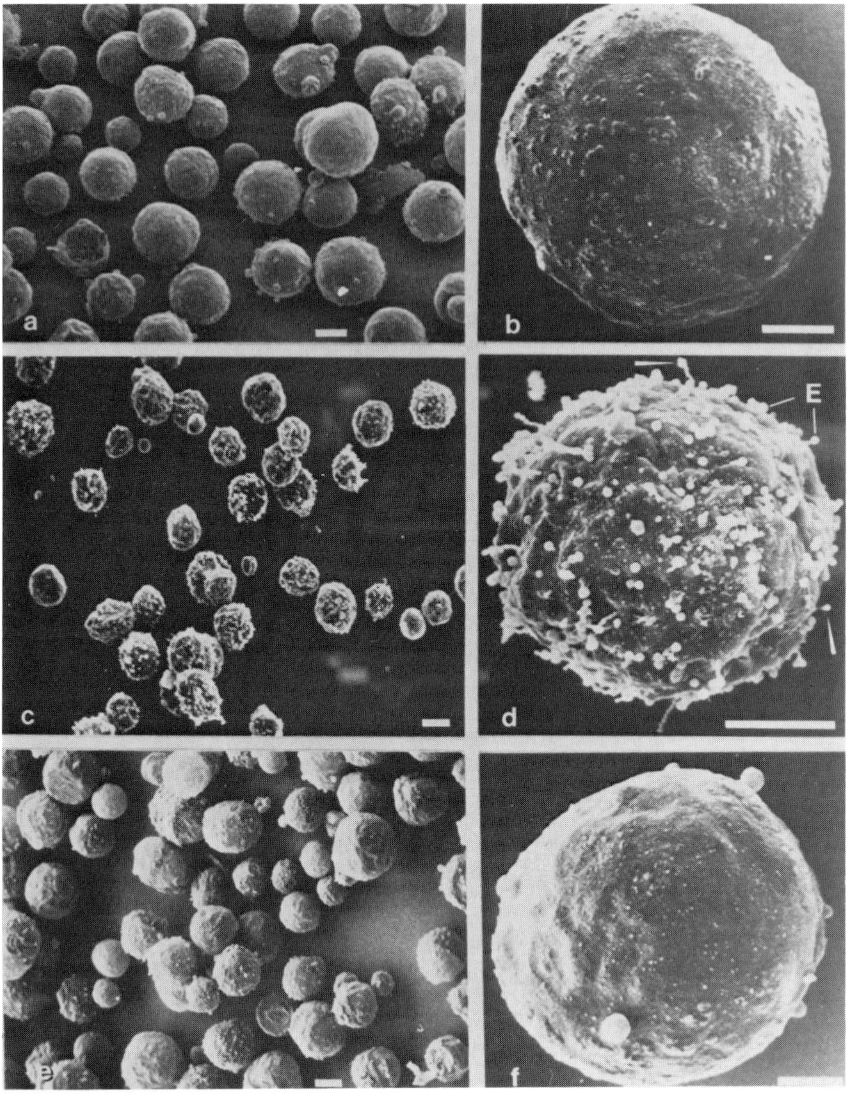

Fig. 6. Scanning electron micrographs of protoplasts isolated from cold-acclimated rye leaves. In (a) and (b), protoplasts were suspended in isotonic (1.03 osmolal) sorbitol solution; the surface of the protoplast is generally smooth. In (c) and (d), the protoplasts were suspended in a hypertonic (2.54 osmolal) sorbitol solution; numerous extrusions are present on the surface of the protoplasts. In (e) and (f), the protoplasts were first contracted in a hypertonic (2.54 osmolal) sorbitol solution and

Behaviour of liposomes during osmotic contraction/ expansion

There are several possibilities to account for the differential behaviour of the plasma membrane during osmotic contraction of protoplasts isolated from non-acclimated leaves in comparison to that of protoplasts isolated from cold-acclimated leaves. These include alterations in the composition of the plasma membrane (either in the lipids or proteins), alterations in the transmembrane distribution of these components, alterations in cytoskeleton–membrane interactions, or possibly even alterations in the cytosolic solutes. To test the hypothesis that the differential behaviour was a consequence of alterations in the plasma membrane lipid composition, computer-enhanced video-microscopy was used to study the behaviour of liposomes formed from the plasma membrane lipids (Steponkus & Lynch 1989*b*). For these studies, large unilamellar vesicles (>20 μm in diameter) were prepared from the total lipid extract of a plasma membrane fraction isolated from either non-acclimated or cold-acclimated rye leaves and are referred to as NA LUVs and ACC LUVs, respectively (Fig. 9). In the initial stages of osmotic contraction both types of liposome behave similarly; the bilayer begins to flutter and the liposome undergoes shape instabilities (i.e. departure from sphericity). Subsequently in NA LUVs, numerous daughter vesicles were subduced from the bilayer and sequestered in the interior of the mother liposome. Following osmotic equilibration, there was good quantitative agreement between the decrease in the surface area of the mother liposome and the total surface area of the daughter vesicles. This behaviour is similar to that observed for LUVs composed of egg lecithin (Boroske *et al.* 1981). Upon dilution of the suspending medium, NA LUVs began to expand osmotically, but lysed before regaining their initial size. In some instances, lysis occurred as a single lytic event without resealing of the bilayer and resulted in the total disintegration of the liposome. In other instances, lysis occurred as a series of transient lytic events during which time there was a transient release of the intraliposomal solution. As a result, there was a progressive diminution in the size of the liposome. Interestingly, the transient lytic event appeared to occur in the same

---

Fig. 6 (*contd*)

then expanded in a more dilute (1.28 osmolal) sorbitol solution; the surface of the protoplast is smooth and free of extrusions. The bars represent 10 μm in (a), (c), and (e), and 5 μm in (b), (d), and (f). (From Gordon-Kamm & Steponkus 1984*b*.)

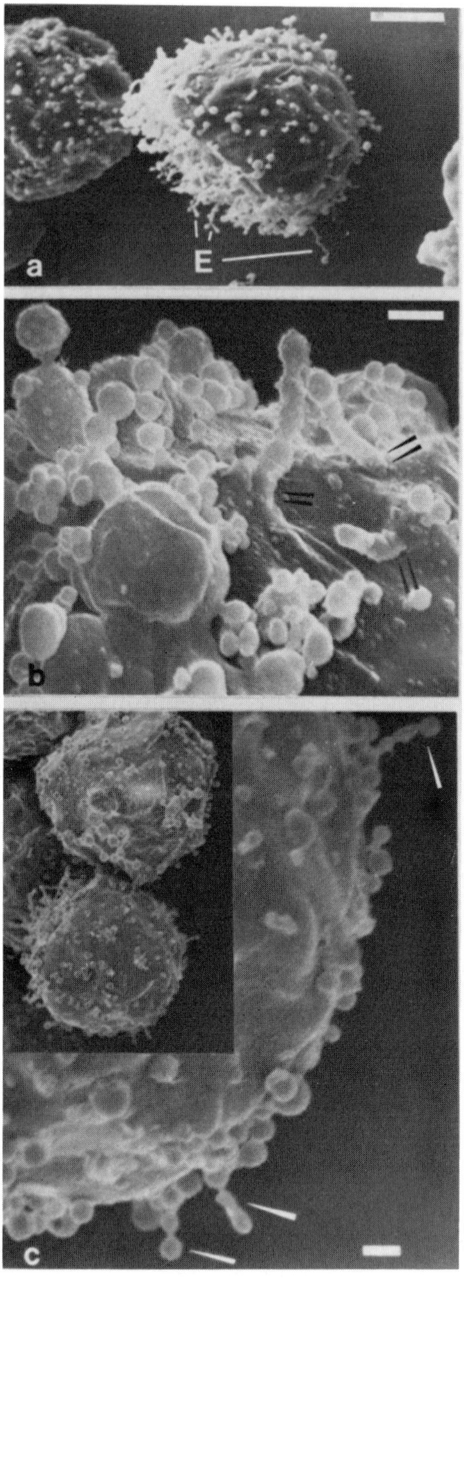
a
E
b
c

location of the bilayer. In one liposome, nine successive lytic events occurred at the same location. This series of transient lytic events has also been observed during the osmotic expansion of LUVs prepared from egg lecithin (Reeves & Dowben 1969, 1970; Callow & McGrath 1985). We interpret the lytic event to result from the increase in tension in the bilayer resulting from the influx of water. However, the increase in area is limited to very small elastic changes before the tension reaches a critical level and results in mechanical failure. The small changes in area are consistent with a k_A on the order of 140 mN m^{-1} as measured for lipid bilayers (Kwok & Evans 1981). Apparently, larger, non-elastic increases in area, which require the incorporation of additional membrane material into the bilayer, are not possible in liposomes.

Osmotic contraction of ACC LUVs resulted in the formation of tubular extrusions or vesicles that remained attached to the mother vesicle (Fig. 10). During subsequent osmotic expansion, the extrusions/vesicles were drawn back into the plane of the bilayer. Although there was a substantial increase in their size, only a few of the LUVs regained their initial size. Nevertheless, these studies demonstrate that the differential behaviour of the plasma membrane during osmotic contraction of NA and ACC protoplasts is also paralleled in liposomes prepared from the respective plasma membrane lipid extracts. Thus, the differential behaviour is, in large part, a consequence of alterations in the lipid composition following cold acclimation.

Membrane engineering studies

Cold acclimation of winter rye results in numerous changes in the lipid composition of the plasma membrane (Lynch & Steponkus 1987). These changes include increases in the proportions of free sterols and phospholipids and decreases in the proportions of glucocerebrosides, sterylglucosides, and acylated sterylglucosides. Phosphatidylcholine and phosphatidylethanolamine are the major phospholipids, and following cold acclimation there is an increase in the proportion of unsaturated

---

Fig. 7. High-magnification scanning electron micrographs of protoplasts that were isolated from cold-acclimated rye leaves and then suspended in a 2.54 osmolal sorbitol solution. Numerous spherical extrusions appear to be sessile on the surface of the protoplasts; others are attached to the surface by tethers (E). Some of the extrusions appear to contain tandem spheres (arrows). In (b) the continuity of the tethers with the plasma membrane is shown (double arrows). The bars represent: (a) 5 μm, (b) 1 μm, (c) 1 μm. (From Gordon-Kamm & Steponkus 1984*b*.)

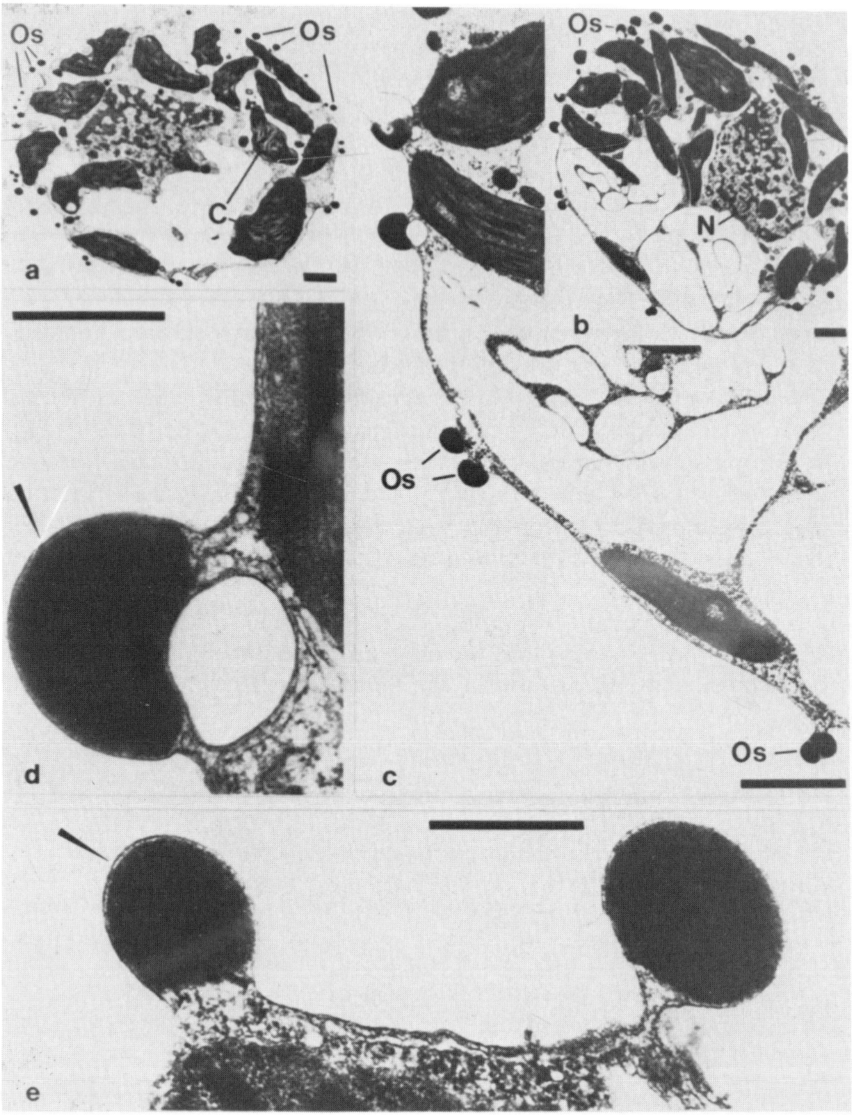

Fig. 8. (a)–(e) Transmission electron micrographs of protoplasts that were isolated from cold-acclimated rye leaves and suspended in a hypertonic (2.54 osmolal) sorbitol solution. Thin sections through the extrusions reveal that the plasma membrane (arrow) is continuous over the extrusions (Os), which have a densely osmiophilic interior, and that the cytoplasm extends into the neck of the extrusions. N, nucleus. The bars represent 0.5 μm. (From Gordon-Kamm & Steponkus 1984*b*.)

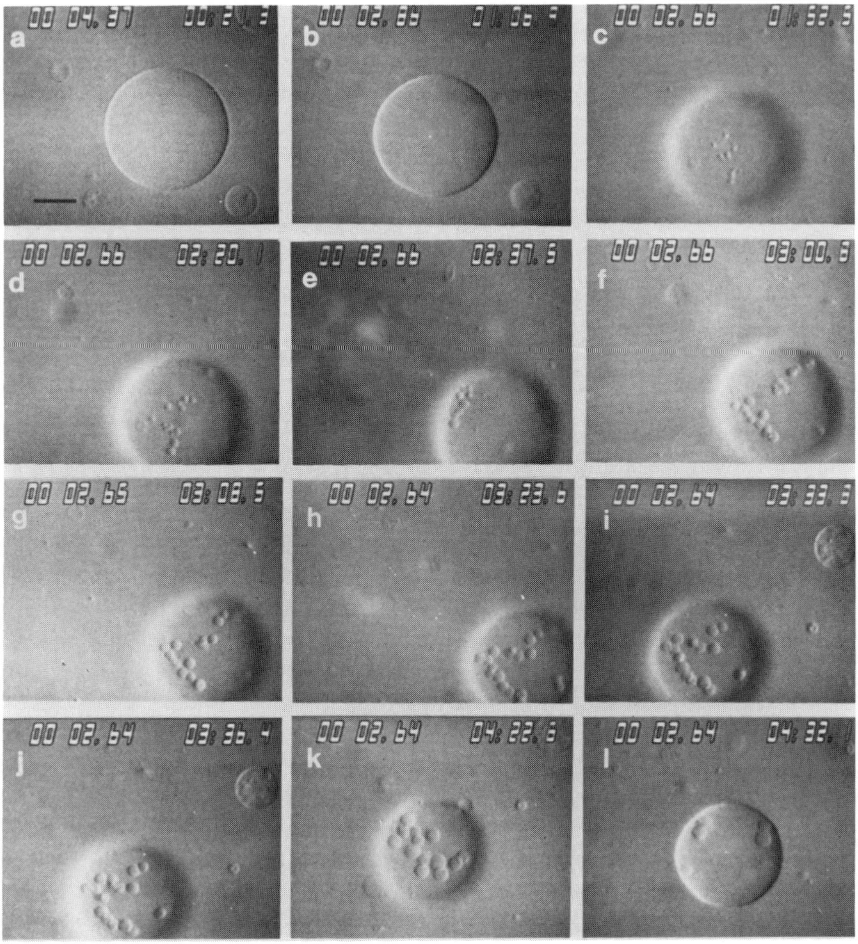

Fig. 9. (a)–(l) Formation of endocytotic vesicles during osmotic contraction of a large unilamellar vesicle (initial diameter of 28 μm) prepared from the plasma membrane lipid extract of non-acclimated rye leaves. The liposome was observed using computer-enhanced video-microscopy with differential interference optics. The bar represents 10 μm. (From Steponkus & Lynch 1989*b*.)

species of both. Although over 100 different lipid molecular species have been identified in the plasma membrane of rye, none of the species is unique to the plasma membrane of either non-acclimated or cold-acclimated leaves. Thus, the differential cryobehaviour of the plasma membrane and liposomes prepared from the plasma membrane lipids is a

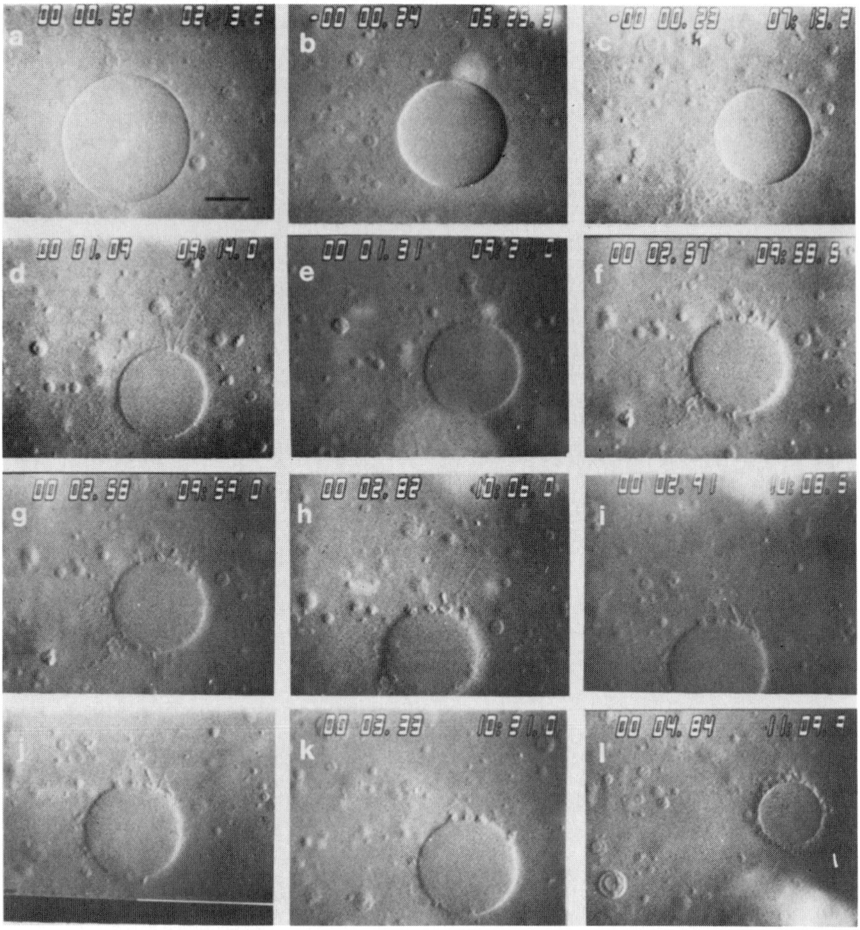

Fig. 10. Formation external extrusions/vesicles during osmotic contraction of a large unilamellar vesicle (initial diameter of 30 μm) prepared from the plasma membrane lipid extract of cold-acclimated rye leaves. The liposome was observed using computer-enhanced video-microscopy with differential interference optics. The bar represents 10 μm. (From Steponkus & Lynch 1989*b*.)

consequence of altered lipid–lipid interactions resulting from changes in the proportions of the individual lipid species. Although the studies of the osmotic behaviour of liposomes composed of the total lipid extract of the plasma membrane established that the differential behaviour of the plasma membrane is a consequence of alterations in the lipid composition of the plasma membrane, such studies did not resolve which of the many changes are responsible for the differential behaviour.

Structure–function relationships responsible for the differential behaviour of the plasma membrane have been established by selective modification of the plasma membrane lipid composition (Steponkus *et al.* 1988). A protoplast–liposome fusion technique (Arvinte & Steponkus 1988) has been used to enrich the plasma membrane with specific lipids. For protoplasts isolated from non-acclimated rye leaves, enrichment of the plasma membrane with either mono- or di-unsaturated species of phosphatidylcholine (1-palmitoyl-2-oleoylphosphatidylcholine, 1-palmitoyl-2-linoleoylphosphatidylcholine, dioleoylphosphatidylcholine, dilinoleoylphosphatidylcholine, or dilinolenoylphosphatidylcholine) transformed the behaviour of the plasma membrane such that during osmotic contraction it behaved as the plasma membrane of protoplasts isolated from cold-acclimated leaves. Following enrichment of the plasma membrane with these lipid species, the plasma membrane formed exocytotic extrusions during osmotic contraction and there was an increase in the tolerance to osmotic excursions (Fig. 11). Dilinoleoylphosphatidylcholine and dilinolenoylphosphatidylcholine were the most effective, whereas fusion with saturated species of phosphatidylcholine (dipalmitoylphosphatidylcholine or dimyristoylphosphatidylcholine) had no effect.

In scanning electron micrographs of protoplasts enriched with dilinoleoylphosphatidylcholine, the morphology of the extrusions was virtually indistinguishable from the morphology of the extrusions formed by protoplasts isolated from cold-acclimated leaves (Fig. 12). To establish a causal relationship between the formation of exocytotic extrusions and the reversibility of osmotic contraction, the expansion potential of the protoplasts was determined during osmotic expansion from both isotonic and hypertonic solutions (Fig. 13). When expanded from an isotonic solution, there was no difference in survival between the untreated protoplasts and those fused with dilinoleoylphosphatidylcholine; i.e. in both, survival declined to 50% after a 25% increase in the surface area of the protoplasts. In contrast, there was a significant difference in survival after expansion from a hypertonic (0.97 osmolal) solution. Survival of the untreated protoplasts decreased to 50% after a 25% increase in the surface area, whereas survival of the protoplasts fused with liposomes of dilinoleoylphosphatidylcholine did not decline appreciably until after the protoplasts were expanded beyond their initial isotonic volume and then was similar to that of protoplasts expanded from isotonic solutions. These results establish that osmotic contraction of protoplasts fused with dilinoleoylphosphatidylcholine is readily reversible because of the

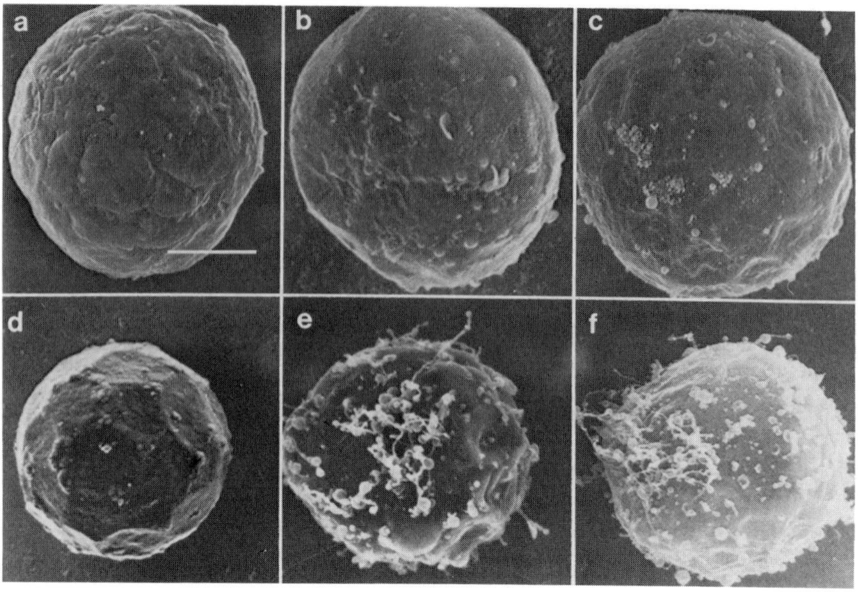

Fig. 11. Scanning electron micrographs of protoplasts isolated from non-acclimated rye leaves and equilibrated in either (a) an isotonic (0.53 osmolal) or (d) a hypertonic (1.0 osmolal) sorbitol solution; protoplasts isolated from non-acclimated rye leaves and fused with liposomes composed of dilinoleoylphosphatidylcholine before equilibration in either (b) an isotonic (0.53 osmolal) or (e) a hypertonic (1.0 osmolal) sorbitol solution; and protoplasts isolated from cold-acclimated rye leaves and suspended in either (c) an isotonic (1.03 osmolal) or (f) a hypertonic (2.0 osmolal) sorbitol solution. The bar represents 10 μm. (From Steponkus *et al.* 1988.)

formation of exocytotic extrusions during osmotic contraction. The fact that the expansion potential from isotonic solutions was not altered indicates that the increased tolerance to osmotic excursions was not the result of excess membrane area resulting from fusion.

Conclusions

These studies demonstrate the dynamic nature of the plasma membrane during osmotic excursions. Although the studies were directed to an understanding of the mechanisms of freezing injury and the effect of cold acclimation, they have provided insights into the intrinsic properties of the plasma membrane and the mechanics of membrane exchange and

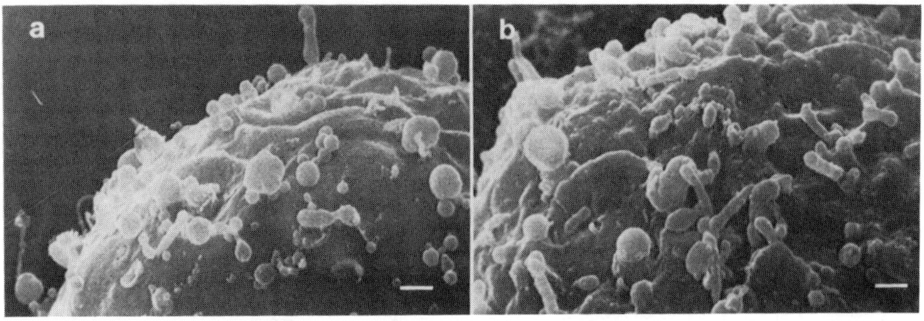

Fig. 12. High-magnification scanning electron micrographs showing the similar morphology of the plasma membrane extrusions formed in (a) protoplasts isolated from cold-acclimated rye leaves and suspended in a hypertonic (2.0 osmolal) sorbitol solution and (b) protoplasts isolated from non-acclimated rye leaves and fused with liposomes of dilinoleoylphosphatidylcholine before suspension in a hypertonic (1.0 osmolal) sorbitol solution. The bars represent 1 μm. (From Steponkus *et al.* 1988.)

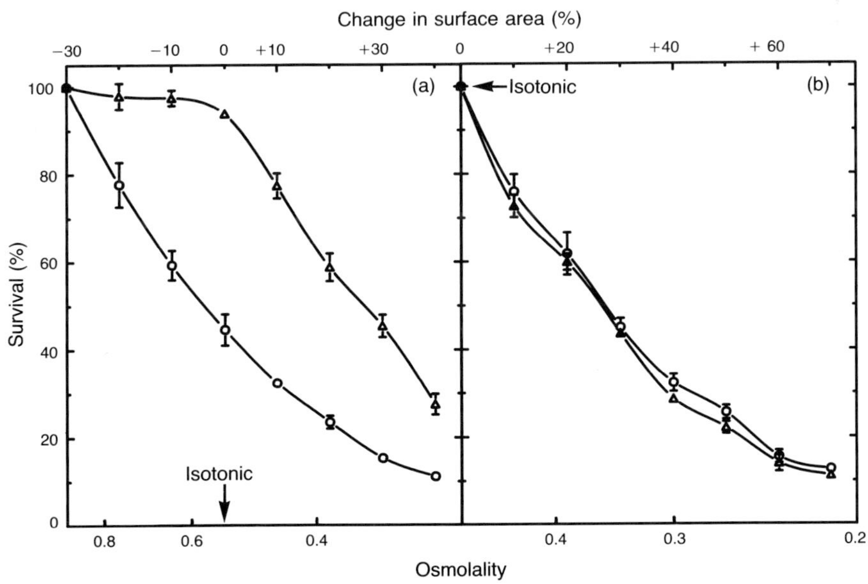

Fig. 13. Survival of protoplasts isolated from non-acclimated rye leaves as a function of the change in the surface area of the plasma membrane during osmotic expansion from either (a) a hypertonic (0.97 osmolal) or (b) an isotonic (0.53 osmolal) sorbitol solution: (○) untreated protoplasts, (△) protoplasts fused with liposomes composed of dilinoleoylphosphatidylcholine. (From Steponkus *et al.* 1988.)

lysis. Nevertheless, although the molecular basis for the altered behaviour following cold acclimation has been identified, the process of endocytotic vesiculation still remains an enigma. However, the commonality of the process in both complex biological membranes and relatively simple lipid vesicles somewhat simplifies studies of this phenomenon.

Acknowledgements

The substantial contributions of several former colleagues including Steven C. Wiest, Michael F. Dowgert, William Gordon-Kamm, Joe Wolfe and Matsuo Uemura, are gratefully acknowledged. Portions of this work were supported by grants from the United States Department of Energy (grant no. DE-FG02-84ER13214) and the United States Department of Agriculture Competitive Grant Program (grant no. 5-CRCR-1-1651 and grant no. 88-37264-3988).

References

Arvinte, T. & Steponkus, P.L. (1988). Characterization of the pH-induced fusion of liposomes with the plasma membrane of rye protoplasts. *Biochemistry*, **27**, 5671–7.

Boroske, E., Elwenspoek, M. & Helfrich, W. (1981). Osmotic shrinkage of giant egg-lecithin vesicles. *Biophysics Journal*, **34**, 95–109.

Callow, R.A. & McGrath, J.J. (1985). Thermodynamic modeling and cryomicroscopy of cell-size, unilamellar, and paucilamellar liposomes. *Cryobiology*, **22**, 251–67.

Dowgert, M.F. & Steponkus, P.L. (1983). Effect of cold acclimation on intracellular ice formation in isolated protoplasts. *Plant Physiology*, **72**, 978–88.

Dowgert, M.F. & Steponkus, P.L. (1984). Behaviour of the plasma membrane of isolated protoplasts during a freeze-thaw cycle. *Plant Physiology*, **75**, 1139–51.

Dowgert, M.F., Wolfe, J. & Steponkus, P.L. (1987). The mechanics of injury to isolated protoplasts following osmotic contraction and expansion. *Plant Physiology*, **83**, 1001–7.

Evans, E.A. & Skalak, R. (1979). *Mechanics and Thermodynamics of Biomembranes*. Boca Raton, FL: CRC Press.

Gordon-Kamm, W.J. & Steponkus, P.L. (1984*a*). The behavior of the plasma membrane following osmotic contraction of isolated protoplasts: implications in freezing injury. *Protoplasma*, **123**, 83–94.

Gordon-Kamm, W.J. & Steponkus, P.L. (1984*b*). The influence of cold acclimation on the behavior of the plasma membrane following osmotic contraction of isolated protoplasts. *Protoplasma*, **123**, 161–73.

Hsiao, T. (1973). Plant responses to water stress. *Annual Review of Plant Physiology*, **24**, 519–70.

Kwok, R. & Evans, E. (1981). Thermoelasticity of large lecithin bilayer vesicles. *Biophysics Journal*, **35**, 637–52.

Lynch, D.V. & Steponkus, P.L. (1987). Plasma membrane lipid alterations associated with cold acclimation of winter rye seedlings (*Secale cereale* L. cv Puma). *Plant Physiology*, **83**, 761–7.

Mitchison, J.M. & Swann, M.M. (1954). The mechanical properties of the cell surface. *Journal of Experimental Biology*, **31**, 443–72.

Reeves, J.P. & Dowben, R.M. (1969). Formation and properties of thin-walled phospholipid vesicles. *Journal of Cell Physiology*, **73**, 49–60.

Reeves, J.P. & Dowben, R.M. (1970). Water permeability of phospholipid vesicles. *Journal of Membrane Biology*, **3**, 123–41.

Scarth, G.W., Levitt, J. & Siminovitch, D. (1940). Plasma-membrane structure in the light of frost-hardening. *Cold Spring Harbor Symposia on Quantitative Biology*, **8**, 102–9.

Steponkus, P.L. (1980). A unified concept of stress in plants? In *Genetic Engineering of Osmoregulation*, ed. D.W. Rains, R.C. Valentine & A. Hollaender, pp. 235–51. New York: Plenum Press.

Steponkus, P.L. (1984). Role of the plasma membrane in freezing injury and cold acclimation. *Annual Review of Plant Physiology*, **35**, 543–84.

Steponkus, P.L., Dowgert, M.F., Evans, R.Y. & Gordon-Kamm, W.J. (1982). Cryobiology of isolated protoplasts. In *Plant Cold Hardiness and Freezing Stress*, vol. 2, ed, P.H. Li & A. Sakai, pp. 459–74. New York: Academic Press.

Steponkus, P.L., Dowgert, M.F. & Gordon-Kamm, W.J. (1983). Destabilization of the plasma membrane of isolated plant protoplasts during a freeze-thaw cycle: the influence of cold acclimation. *Cryobiology*, **20**, 448–65.

Steponkus, P.L. & Lynch, D.V. (1989*a*). Freeze/thaw-induced destabilization of the plasma membrane and the effects of cold acclimation. *Journal of Bioenergetics and Biomembranes*, **21**, 21–41.

Steponkus, P.L. & Lynch, D.V. (1989*b*). The behavior of large unilamellar vesicles of rye plasma membrane lipids during freeze/thaw-induced osmotic excursions. *Cryo-Letters*, **10**, 43–50.

Steponkus, P.L., Uemura, M., Balsamo, R.A., Arvinte, T. & Lynch, D.V. (1988). Transformation of the cryobehavior of rye protoplasts by modification of the plasma membrane lipid composition. *Proceedings of the National Academy of Sciences, USA*, **85**, 9026–30.

Steponkus, P.L. & Wiest, S.C. (1979). Freeze-thaw-induced lesions in the plasma membrane. In *Low Temperature Stress in Crop Plants: The Role of the Membrane*, ed. J.M. Lyons, D. Graham & J.K. Raison, pp. 231–54. New York: Academic Press.

Steponkus, P.L., Wolfe, J. & Dowgert, M.F. (1981). Stresses induced by contraction and expansion during a freeze-thaw cycle: a membrane perspective. In *Effects of Low Temperatures on Biological Membranes*, ed. G.J. Morris & A. Clarke, pp. 307–22. London: Academic Press.

Tornava, S.R. (1939). Expansion capacity of naked protoplasts. *Protoplasma*, **32**, 329–41.

Wiest, S.C. & Steponkus, P.L. (1978). Freeze-thaw injury to isolated spinach protoplasts and its simulation at above freezing temperatures. *Plant Physiology*, **62**, 699–705.

Wolfe, J., Dowgert, M.F., Maier, B. & Steponkus, P.L. (1986*a*). Hydration, dehydration, and the stresses in membranes. In *Membranes, Metabolism, and Dry Organisms*, ed. A.C. Leopold, pp. 286–305. Ithaca, NY: Comstock Press.

Wolfe, J., Dowgert, M.F. & Steponkus, P.L. (1985). Dynamics of membrane exchange of the plasma membrane and the lysis of isolated protoplasts during rapid expansions in area. *Journal of Membrane Biology*, **86**, 127–38.

Wolfe, J., Dowgert, M.F. & Steponkus, P.L. (1986*b*). Mechanical study of the deformation and rupture of the plasma membranes of protoplasts during osmotic expansions. *Journal of Membrane Biology*, **93**, 63–74.

Wolfe, J. & Steponkus, P.L. (1981). The stress–strain relation of the plasma membrane of isolated plant protoplasts. *Biochimica et Biophysica Acta*, **643**, 663–8.

Wolfe, J. & Steponkus, P.L. (1983). Mechanical properties of the plasma membrane of isolated protoplasts. *Plant Physiology*, **71**, 276–85.

M. W. STEER AND D. O'DRISCOLL

Vesicle dynamics and membrane turnover in plant cells

Cellular function depends on the integration of chemical processes supported by structural elements. Transport of certain cellular products is achieved by ensuring that they are retained within a membrane-enclosed system and packaged into moveable entities known as vesicles. These membrane-bounded bodies serve as carriers of discrete packets of product from one membrane compartment to another. In this chapter we concentrate on one class of vesicle, the secretory vesicles formed by the Golgi apparatus. This restriction is imposed by our lack of experimental methods for examining movements of other vesicles, for example coated vesicles. Only when a complete analysis of vesicle traffic in cells is available will it be possible properly to evaluate their contribution to cellular function.

Secretion involves transfer of vesicles from elements of the Golgi apparatus, the dictyosomes, to the cell surface, where they fuse with the plasma membrane. Most of the vesicle membrane fuses with, and is incorporated into, the plasma membrane. The contents are discharged into the extracellular space, which in the case of plant cells is the cell wall.

Among plant cell secretory functions that have been identified, the release of cell wall polysaccharides and other structural cell wall components (e.g. extensin and lignin) appear to be the most universal, along with delivery of new membrane-bound proteins to the cell surface (cellulose-synthesising complexes, Haigler & Brown 1986; ion channels and ATPases). Other functions include the release of mucilages (Rougier 1972), digestive enzymes (Cornejo *et al.* 1988) and wall hydrolases to the outside of the cell, and the delivery of storage proteins to the internal vacuole system (Parker & Hawes 1982; Greenwood & Chrispeels 1985; Krishnan *et al.* 1986). Since cell wall synthesis is intimately connected with cell expansion, and thus plant growth, it is clear that the rate of secretion is an important component affecting plant development (Phillips *et al.* 1988, and references therein). This is confirmed by studies demonstrating that inhibition of secretion prevents cell growth.

As a first step to understanding how the cell controls secretion we need to know how to monitor the rate of this process, and find out to what extent this rate can be regulated by changing cell growth conditions. We also need to ascertain if the process of vesicle formation is driven by product synthesis and accumulation within the dictyosomes. In other words how closely are the packaging and manufacturing processes linked?

At the present time our information on even this major pathway of vesicle traffic is fragmentary. In this chapter we first review the methods available for estimating the rate of vesicle movement via this pathway. Information on variation in this rate, induced by varying experimental and growth conditions in one system, the growing pollen tube, will be discussed. The rate of secretion in cells is examined in a typical growing plant tissue, the extending oat coleoptile, and in the non-expanding secretory cells of the root cap. These results are brought together to review the overall secretory process, first at the level of the dictyosome and dictyosome activity, then at the level of vesicle function, and finally in terms of the capacity of the cytoplasm to support the enclosing cell plasma membrane.

Despite the absence of direct methods for analysing another major vesicle flow pathway in plants, endocytosis, the evidence from secretion rates in plant cells and from endocytotic rates in animal cells is combined to provide some insights into this process.

Methods for determining secretory vesicle traffic rates

The first estimate of secretory vesicle traffic rates in plant cells was made by Schnepf in 1961. He observed the rate of increase in the size of the mucilage droplet on the surfaces of the stalked glands of *Drosophyllum*, and so obtained an estimate of exudation rate per unit area (μm^2) of epidermal cell surface. Microscopy was used to find the average diameter of the secretory vesicles, and the steady-state population of vesicles per square micrometre of cell surface. From these data he was able to determine the secretion rate of vesicles and their transit time in the cytoplasm.

In essence, Schnepf was evoking the general principle that a unit volume of secretion on the outside of the cell surface originates from a unit volume of material delivered to the surface by secretory vesicles.

Later, the same principle was used by Morré & van der Woude (1974) in their estimation of the vesicle production rates of pollen tubes. The rate of new cell wall material originating at the growing tip was estimated from the tube growth rate and tube wall diameter. This secreted wall volume was assumed to be equivalent to the same volume of vesicle contents discharged at the tip. The mean vesicle diameter was measured and an estimate made of the rate of vesicle secretion (1075 vesicles min^{-1}). Coincidentally this rate of secretion provided all the additional membrane required to augment the plasma membrane surface during tube extension.

This approach to the determination of vesicle secretion rates depends on an equivalence between the volume of secreted material, recorded from the surface of a gland or organ, and the volume of material delivered by the vesicles of the underlying cells. In fact, such an equivalence is rarely, if ever, encountered. Other factors intervene that either increase or decrease the volume of secreted material. In the two examples discussed here, osmotic swelling of mucilage (with water derived from the tissue) and closer packing of polysaccharide in the cell wall than in the vesicles lead, respectively, to over- and underestimation of the true rate of vesicle secretion.

A different approach was adopted by Picton & Steer (1981). They used cytochalasins to block movement of secretory vesicles to the cell surface, following a report by Pope *et al.* (1979) that, in plant cells, vesicles accumulate in the cytoplasm when the tissue is exposed to these drugs. Estimates were made of vesicle populations in control cells and in cells that had been treated with cytochalasin D for a measured period of time before glutaraldehyde fixation. These estimates were based on stereological analysis of electron micrographs from thin sections of the cells, and yielded the vesicle density (number per unit volume of cytoplasm). Vesicle production rate per unit volume of cytoplasm was determined by subtracting the control value from the density of vesicles in treated cells, and dividing the result by the length of time that the cells were exposed to the drug (Fig. 1). By also determining the density of dictyosomes in the cytoplasm, it was possible to make an estimate of dictyosome activity (vesicles produced per dictyosome per minute).

At first this method appeared to give satisfactory results, enabling production rates to be estimated in pollen tubes (Fig. 2), growing normally and under experimental conditions (as a result of inhibiting protein synthesis with cycloheximide, Picton & Steer 1983*b*). This demonstrated that it was possible to distinguish between dynamic and static populations of vesicles in cells. Later work on root cap cells necessitated longer exposure to cytochalasin as this tissue has a lower rate of vesicle production, so more time is required to accumulate a significant increase in the vesicle numbers (Shannon *et al.* 1984). The rate of production was found to fall to about 10% of the initial rate after 2 h, so it was concluded that the drug treatment leads to a progressive inhibition of vesicle production.

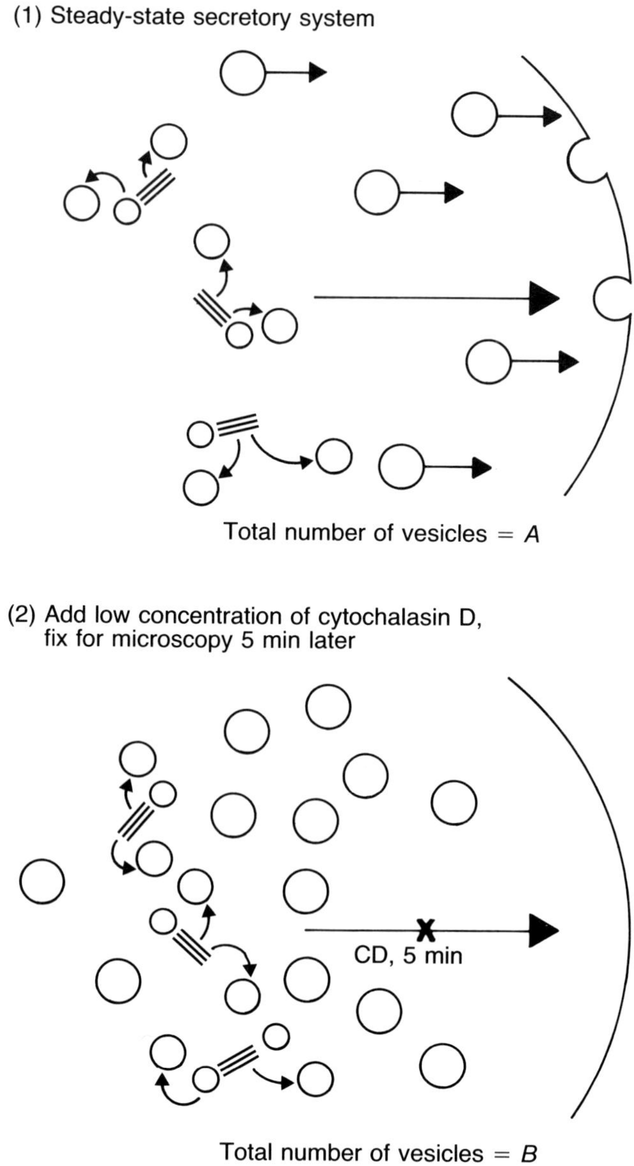

Production = $B-A$

Production rate = $B-A/5\ \text{min}^{-1}$

Production rate per dictyosome = $(B-A/5)/3\ \text{min}^{-1}$

The reasons for this inhibition are not clear. Assuming that cytochalasin is only acting on the cell's cytoskeleton, inhibition of production must be due to an interruption in the supply of essential components for either, or both, the vesicle contents and vesicle membranes. These originate from the endoplasmic reticulum, the Golgi apparatus itself, and also from components of the plasma membrane recycled by endocytosis. It seems most likely that, as exocytosis is inhibited, endocytosis will be considerably curtailed, removing one source of macromolecules for the synthesis of new vesicles.

The results obtained for vesicle production rates using the cytochalasin method appear to give a good indication of the actual rates, since inhibition of vesicle fusion at the plasma membrane leads to an accumulation of vesicles at a similar rate in pollen tubes (Picton & Steer 1985).

Vesicle production rates in pollen tubes

Germinating pollen grains, at least bicellular ones, have highly active Golgi that result in a massive formation of vesicles and rapid tube growth. The early studies of Morré & van der Woude (1974) demonstrated the scale of this activity, with at least 1000 vesicles produced per minute by each lily pollen tube. The improved methods developed by Picton & Steer (1983*a*) raised this value to 3000–5000 vesicles per minute in *Tradescantia*, with each dictyosome producing approximately two vesicles per minute.

While some variation in vesicle formation rates was recorded between different experiments on pollen of the same species (*Tradescantia*), it was not possible to demonstrate any link between vesicle production rate and pollen tube growth rate. A four-fold change in tube growth rate, achieved by altering the external calcium concentration, did not lead to any change in the rate of vesicle formation (Picton & Steer 1983*a*).

Dependence of vesicle formation on continued protein synthesis was

---

Fig. 1. The cytochalasin-based method for estimating secretory vesicle formation rates. (1) The system under investigation must be producing secretory vesicles at a steady rate, with consistent movement to, and fusion with, the plasma membrane. (2) Under these conditions, application of cytochalasin will result in inhibition of transport to the cell surface without, in the short term, interrupting vesicle formation. Vesicles accumulate in the cytoplasm until the cell is fixed for electron microscopy. Vesicle numbers, and other numerical and volume determinations, must be made using appropriate quantitative methods. CD, cytochalasin D.

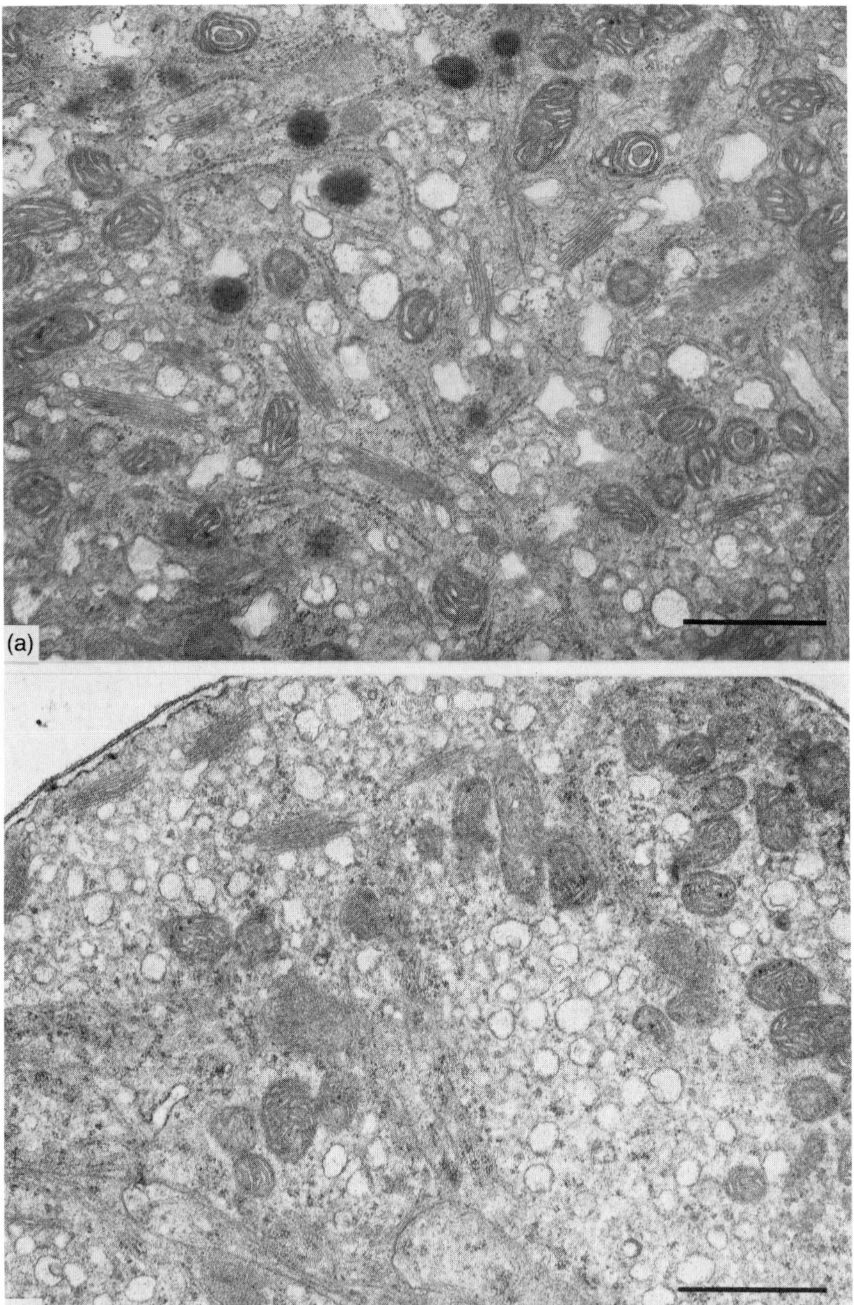
(a)
(b)

demonstrated by exposing growing tubes to cycloheximide for varying periods of time before assessing vesicle production rates with cytochalasin (Picton & Steer 1983*b*). After 30 min in cycloheximide the tubes contained more vesicles per unit volume than controls, but these were shown to be a static population that was not being added to by vesicle formation, nor being depleted by exocytosis. Precisely why these vesicles are not secreted was not determined, but it is possible that their synthesis under cycloheximide inhibition had deprived them of an essential protein for transport to the growing tip.

The information on normal pollen tube growth implies that a considerable proportion of the plasma membrane delivered to the tube tip is recycled (Picton & Steer 1983*a*). Unfortunately there is no information on corresponding coated vesicle traffic from the cell surface. Coated pits are encountered only rarely on the surface plasma membrane. Coated vesicle populations are predominantly located around the Golgi apparatus. This population, observed in electron micrographs after standard fixation, is larger in cycloheximide-treated pollen tubes than in controls (Picton 1981). This information does not, of course, allow any statement to be made about the flow of coated vesicles through this population in either condition.

Secretion in plant tissue cells

Compared with pollen tubes, vesicle formation rates are more difficult to determine in cells of plant tissues. This is due to problems of cytochalasin access to the cells and the generally lower rates of vesicle production in these cells. Low production rates are difficult to detect by the cytochalasin method, due to the inherent limitations imposed by the stereological methods for estimating vesicle numbers from electron micrographs.

Gland cells in the ovary of *Aptenia cordifolia* release mucilage as the result of dictyosome secretory activity. Using the cytochalasin method, Kristen & Lockhausen (1983) found that the dictyosomes form vesicles at a rate of about one per minute. The vesicles supply enough new membrane to the cell surface to turn over the entire plasma membrane once every 10 min. The residence time of the vesicles was found to be about 9 min.

Fig. 2. Micrographs of *Tradescantia* pollen tubes (a) before and (b) after treatment with cytochalasin D for 10 min. Note the incresed number of secretory vesicle profiles in the cytoplasm of the treated tube. The bar represents 1 μm.

Root cap cells of *Zea mays* produce a mucilage slime, synthesised in the dictyosomes. Rates of vesicle formation in these cells (0.4 dictyosome^{-1} min^{-1}), although lower than in pollen tubes, are still sufficient to turn over the entire plasma membrane of the cell once every 10 min. The residence time for these vesicles, the average time from release by the dictyosomes to fusion with the plasma membrane, is 18 min (Shannon & Steer 1984).

Oat coleoptile cells undergo rapid elongation in the dark, maintaining their plasma membrane integrity and cell wall thickness (Quaite *et al.* 1983). The synthetic activity necessary to sustain these processes is provided by a very thin layer of cytoplasm, lying between the plasma membrane and the tonoplast, that encloses a large central vacuole. Small dictyosomes form secretory vesicles at such a high rate that they provide an excess of new membrane to the rapidly expanding cell surface, so that about one-third of the membrane delivered is recycled back into the cell. Vesicles are produced at the rate of 3 dictyosome^{-1} min^{-1} and have a resident time of about 17 min in the cytoplasm. Overall, each cell produces 40 000–50 000 vesicles min^{-1} (Phillips *et al.* 1988).

Dictyosome activity

Dictyosome cisternae house enzyme systems for polysaccharide synthesis (glycosyltransferases) and are involved in sorting membrane and cisternal products into secretory vesicles. A major avenue of inquiry is into the control of dictyosome activity, since cell growth depends on new plasma membrane and cell wall formation. Inhibition of secretion inhibits cell growth, both when vesicle transport is inhibited by cytochalasins and when dictyosome function is inhibited by the ionophore monensin, and the protein synthesis inhibitor cycloheximide.

Since the size of dictyosomes varies considerably between plant tissues, direct comparison of vesicle formation rates does not give an indication of relative levels of dictyosome activity between different tissues. Expressing activity in terms of a unit volume of dictyosome provides one basis for comparison. Since vesicle sizes vary considerably as well, output can be compared more effectively by looking at the production rate of a unit area of vesicle membrane (Table 1). This analysis shows that pollen tubes have the most active dictyosomes, producing 3.6 (μm membrane)^2 (μm dictyosome)$^{-3}$ min^{-1}, with coleoptile epidermal cells producing 1.8 μm^2 μm^{-3} min^{-1} and root cap cells 0.65 μm^2 μm^{-3} min^{-1} (Steer 1988).

These differences in levels of activity partly reflect the ability of dictyosomes to partition activity between satisfying cell surface requirements for new membrane and providing materials for extracellular function.

Table 1. *Dictyosome activity*

Cell type	Dictysome volume (μm^3)	Area of vesicle membrane (μm^2 (μm dictyosome)$^{-3}$ min^{-1})
Root cap	0.3	0.65
Coleoptile epidermis	0.03	1.8
Pollen tube	0.03	3.6

Table 2. *Vesicle size and vesicle function*

Cell type	Vesicle diameter (μm)	Dictyosome volume (μm^3)	Vesicle function
Root cap	0.4	0.3	Mucilage secretion
Coleoptile	0.07	0.03	PM and wall synthesis
Pollen tube	0.2	0.03	PM and wall synthesis

Note: PM, plasma membrane.

The large secretory vesicles produced by root cap cells only deliver about 50% of membrane area per unit volume of contents compared with vesicles from pollen tubes and 18% when compared with vesicles from epidermal cell dictyosomes. Since root cap cells are mature and have ceased to enlarge, there is no requirement for additional membrane, unlike the other two cell types, which are undergoing rapid expansion. So it appears that dictyosome activity is shifted towards production of root cap mucilages packaged into larger vesicles.

Control of vesicle size provides a simple physical basis for adjusting these requirements of dictyosome function. More subtle controls must also exist. For example, the vesicles of pollen tubes deliver essentially all the carbohydrates required for cell wall formation at the tube tip. These vesicles are about three times the diameter of coleoptile epidermal cell vesicles, which, while also concerned with delivery of membrane, deliver only a proportion of the wall polysaccharide. The remainder of the epidermal cell wall is synthesised at the cell surface, by plasma membrane-bound cellulose synthases (provided by exocytosis). The epidermal cell vesicles deliver three times as much membrane per unit volume of vesicle, as do their counterparts in pollen tubes (Table 2).

Epidermal cells probably contain the minimal cytoplasmic requirement for maintenance of cell integrity and growth potential. In these cells the

Table 3. *Plasma membrane turnover as a function of cytoplasmic capacity*

Cell type	Volume of cytoplasm available (μm^3 (μ PM)$^{-2}$)	PM turnover times (min)
Root cap	5	10
Coleoptile epidermis	0.4	200

Note: PM, plasma membrane.

average volume of cytoplasm available for the maintenance of each square micrometre of plasma membrane is only 0.4 μm^3 compared with 5 μm^3 for each square micrometre of plasma membrane in a root cap cell. Even so the epidermal cell dictyosomes still produce more plasma membrane than is needed for cell extension (Table 3).

Endocytosis

Endocytosis, involving the formation of coated vesicles from coated pits on the plasma membrane, is presumed to recycle all the excess membrane delivered to the cell surface by secretory vesicles. Measurement of the actual rate of endocytosis in plant cells has not yet been achieved. This is in marked contrast to work on animal cells, where the rate of endocytotic coated vesicle formation can be estimated by counting the number of labelled vesicles appearing in the cytoplasm after a known time period of exposure to a cell surface label, such as ferritin or colloidal gold. In plant cells the cell wall acts as an effective barrier to these labels, because they are larger than the largest size of particles that are able to pass through the wall (10–12 nm compared with 3–6 nm).

Experiments with protoplasts have confirmed that endocytotic events in plants are, at least morphologically, similar to those in animals (see Fowke *et al.*, this volume). Work with gold particles in the 2–7 nm size range has shown that such particles can cross the cell wall and be taken up into membrane-bounded compartments within the cytoplasm. However, the rate of penetration of the wall by these particles is very slow, so that they cannot be used to provide good estimates of the endocytotic rate (O'Driscoll & Steer 1990).

Morphological observations of the density of coated pits on the plasma membrane were made on dry cleaved preparations of *Equisetum* root hairs (Emons & Traas 1986). It was found that actively growing cells have a higher density (8 μm^{-2}) of coated pits than mature cells (2–4 μm^{-2}). Using literature values for the lifetime of a coated pit (30–60 s), Emons &

Traas were able to estimate the time it would take to internalise an area equivalent to the whole cell plasma membrane. For the growing cells this turnover time was estimated to be 20–40 min, while for the mature cells it was much longer, 60–115 min. These estimates span the range of turnover times for the plasma membranes proposed on the basis of secretory vesicle production rates.

The rate of endocytosis, based on the assumption that all plasma membrane recycling is channelled through this route, is high, with up to 30 000 vesicles per minute being formed by a pollen tube, or 14 000 vesicles per minute being formed by an epidermal cell. From these figures and the earlier data presented here on rates of secretory vesicle formation, it is clear that plant cells exhibit a very active vesicle traffic, requiring considerable ultrastructural and biochemical organisation. Our understanding of the significance of this activity in plants is far from complete. Some of the functions served by this traffic have been outlined, but only in a very general sense, so we await with interest the results of future detailed analyses of these events.

Conclusion

The data discussed in this chapter serve to demonstrate amply the high level of vesicle traffic in plant cells. By analogy, we can assume that similar levels of activity are also found in animal cells, although quantitative experimental evidence for this is still awaited. This information represents only a small beginning to our understanding of the control mechanisms involved. The close link between dictyosome activity and cell growth, found in plant tissue cells, indicates that multiple cellular activities can be induced to operate in a concerted manner by external conditions. These activities include both secretory and retrieval mechanisms.

The results reviewed here on two of the vesicle pathways found in plant cells have implications for the level of operation of a third pathway, that from endoplasmic reticulum to the Golgi apparatus. As emphasised by Robinson & Kristen (1982), the flow of vesicles between these two compartments is less readily seen in electron micrographs of plant cells than in correspondingly active animal cells (exceptions are: pollen tubes, Picton & Steer 1983*b*; Steer & Steer 1989; and algal cells, see Gunning & Steer 1975, for example). If all of the secretory vesicle membrane originated from the endoplasmic reticulum, then there should be a corresponding flow of transitional vesicles to the *cis*-faces of the dictyosomes. The apparent lack of an equivalent flow in plant cells may mean that the estimated high levels of endocytosis serve to recycle membrane components to the dictyosome cisternae, as has been demonstrated in animal cell systems (Woods *et al.* 1986). The endoplasmic reticulum would, on this basis, be viewed as an organelle that is only concerned with supplying the additional secretory vesicle membrane required for plasma membrane expansion during cell growth.

The interface between the inside of the cell wall and the plasma membrane is now revealed as a very dynamic environment, with rapid and continuous changes in the plasma membrane occurring at the same time as polysaccharide delivery and synthesis for expanding cell wall layers. Although some consideration has been given to these changes (Schnepf *et al.* 1985; Steer 1988) much remains to be done to link physiological, biochemical and structural information on plant cell plasma membrane activities with its associated exocytotic and endocytotic vesicle traffic.

References

Cornejo, M.J., Platt-Aloia, K.A., Thomson, W.W. & Jones, R.L. (1988). Effects of GA_3 and Ca^{2+} on barley aleurone protoplasts: a freeze-fracture study. *Protoplasma*, **146**, 157–65.

Emons, A.M.C. & Traas, J.A. (1986). Coated pits and coated vesicles on the plasma membrane of plant cells. *European Journal of Cell Biology*, **41**, 57–64.

Greenwood, J.S. & Chrispeels, M.J. (1985). Immunocytochemical localisation of phaseolin and phytohemagglutinin in the endoplasmic reticulum and Golgi complex of developing bean cotyledons. *Planta*, **164**, 295–302.

Gunning, B.E.S. & Steer, M.W. (1975). *Ultrastructure and the Biology of Plant Cells*. London: Arnold.

Haigler, C.H. & Brown, R.M. (1986). Transport of rosettes from the Golgi apparatus to the plasma membrane in isolated mesophyll cells of *Zinnia elegans* during differentiation to tracheary elements in suspension cultures. *Protoplasma*, **134**, 111–20.

Krishnan, H.B., Franceschi, V.R. & Okita, T.W. (1986). Immunochemical studies on the role of the Golgi complex in protein-body formation in rice seeds. *Planta*, **169**, 471–80.

Kristen, U. & Lockhausen, J. (1983). Estimation of Golgi membrane flow rates in ovary glands of *Aptenia cordifolia* using cytochalasin B. *European Journal of Cell Biology*, **29**, 262–7.

Morré, D.J. & van der Woude, W.J. (1974). Origin and growth of cell surface components. In *Macromolecules Regulating Growth and Development*. II. *Formation and Organisation of Plant Cell Walls and the Plasma Membrane*, ed. E.D. Hay, T.S. King & J. Papaconstantinou, pp. 81–111. New York: Academic Press.

O'Driscoll, D. & Steer, M.W. (1990). The development of a marker for estimating the rate of endocytosis in plant cells. *Cell Biology International Reports*, **14**, 981–8.

Parker, M.L. & Hawes, C.R. (1982). The Golgi apparatus in developing endosperm of wheat (*Triticum aestivum* L.). *Planta*, **154**, 277–83.

Phillips, G.D., Preshaw, C. & Steer, M.W. (1988). Dictyosome vesicle production and plasma membrane turnover in auxin-stimulated outer epidermal cells of coleoptile segments from *Avena sativa* (L.). *Protoplasma*, **145**, 59–65.

Picton, J.M. (1981). *Quantitative ultrastructural studies of vesicle production and utilisation in pollen tubes.* Ph.D. thesis, The Queen's University of Belfast.

Picton, J.M. & Steer, M.W. (1981). Determination of secretory vesicle production rates by dictyosomes in pollen tubes of *Tradescantia* using cytochalasin D. *Journal of Cell Science*, **49**, 261–72.

Picton, J.M. & Steer, M.W. (1983*a*). Membrane recycling and the control of secretory activity in pollen tubes. *Journal of Cell Science*, **63**, 303–10.

Picton, J.M. & Steer, M.W. (1983*b*). The effect of cycloheximide on dictyosome activity in *Tradescantia* pollen tubes determined using cytochalasin D. *European Journal of Cell Biology*, **29**, 133–8.

Picton, J.M. & Steer, M.W. (1985). The effects of ruthenium red, lanthanum, fluorescein isothiocyanate and trifluoperazine on vesicle transport, vesicle fusion and tip extension in pollen tubes. *Planta*, **163**, 20–6.

Pope, D.G., Thorpe, J.R., Al-Azzawi, M.J. & Hall, J.L. (1979). The effect of cytochalasin B on the rate of growth and ultrastructure of wheat coleoptiles and maize roots. *Planta*, **144**, 373–83.

Quaite, E., Parker, R.E. & Steer, M.W. (1983). Plant cell extension: structural implications for the origin of the plasma membrane. *Plant, Cell and Environment*, **6**, 429–32.

Robinson, D.G. & Kristen, U. (1982). Membrane flow via the Golgi apparatus of higher plant cells. *International Review of Cytology*, **77**, 89–127.

Rougier, M. (1972). Etude cytochimique des squamules d'*Elodea canadensis*. Mise en évidence de leur sécrétion polysaccharidique et de leur activité phosphatasique acide. *Protoplasma*, **74**, 113–31.

Schnepf, E. (1961). Quantitative Zusammenhänge zwischen der Sekretion des Fangschleimes und der Golgi-Struckturen bei *Drosophyllum lusitanicum*. *Zeitschrift für Nturforschung*, **16b**, 605–10.

Schnepf, E., White, O., Rudolph, U., Deichgräber, G. & Reiss, H.-D. (1985). Tip cell growth and the frequency and distribution of particle rosettes in the plasmalemma: experimental studies in *Funaria protonema* cells. *Protoplasma*, **127**, 222–9.

Shannon, T.M., Picton, J.M. & Steer, M.W. (1984). The inhibition of dictyosome vesicle formation in higher plants by cytochalasin D. *European Journal of Cell Biology*, **35**, 144–7.

Shannon, T.M. & Steer, M.W. (1984). The root cap as a test system for the evaluation of Golgi inhibitors. I. Structure and dynamics of the secretory system and response to solvents. *Journal of Experimental Botany*, **35**, 1697–707.

Steer, M.W. (1988). Plasma membrane turnover in plant cells. *Journal of Experimental Botany*, **39**, 987–96.

Steer, M.W. & Steer, J.M. (1989). Pollen tube tip growth. *New Phytologist*, **111**, 323–58.

Woods, J.M., Doriaux, M. & Farquar, M.G. (1986). Transferrin receptors recycle to *cis* as well as *trans* Golgi cisternae in Ig-secretory myeloma cells. *Journal of Cell Biology*, **103**, 277–86.

N. HARRIS AND M.D. WATSON

Vesicle transport to the vacuole and the central role of the Golgi apparatus

Vesicles, vacuoles and the Golgi apparatus are parts of the dynamic continuum of the endomembrane system within plant and yeast cells. The vacuoles can fulfil a number of compartmental functions including storage of both material to be recycled and material to be effectively 'excreted' from the cytosol, breakdown and turnover of organelles and macromolecules, and the osmotic and structural roles related to cell turgor and growth. The vacuole(s) within a single cell may fulfil all of these functions at different stages of the cell cycle.

The Golgi apparatus, with its constituent dictyosomes, is involved in both the formation of vacuoles and transport of material to them, but it may also be involved simultaneously in transport of vesicles and metabolites to other sites such as the plasma membrane.

This chapter considers the structural and functional relationships of the vesicles associated with the Golgi apparatus and their role in the various transport routes to the vacuole (Fig. 1). The various methods used to dissect such relationships are considered and some examples of different patterns of transport in different tissues, and at different stages within the same tissue, are examined.

An overview of yeast studies is included because the unique combinations of methods that can be applied to this organism makes it one of the principal model systems for many aspects of cell biology. Biochemistry, molecular biology, cytochemistry and genetics can be applied in unison, and such combinations are particularly valuable in studies of vesicle transport and protein sorting. The isolation by Novick *et al.* (1980) of mutant strains of yeast which are defective in secretion (*sec* mutations) has opened the way for a thorough molecular dissection of the secretory pathway. Recent studies have shown that some of the proteins involved in sorting are functionally conserved between animals and yeast (Wilson *et al.* 1989), immediately prompting the question of whether or not the conservation also extends to plants. The mechanism of sorting proteins in the Golgi apparatus and their transport to the vacuole in yeast and the

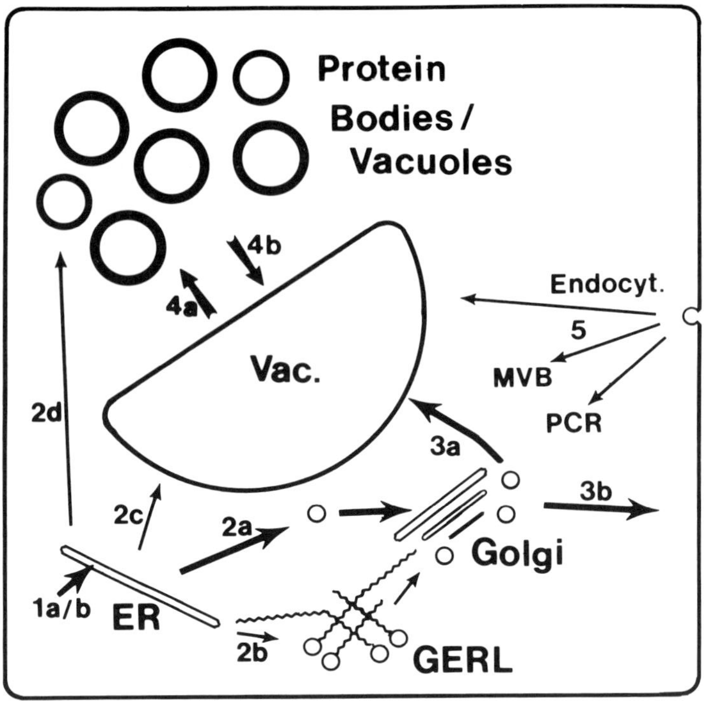

Fig. 1. Diagram of pathways involving vesicular transport, the Golgi apparatus and the vacuole, in plant cells. 1, Transport to (a) ER and (b) ER lumen; 2a, ER to Golgi transport *via* transition vesicles; 2b, ER–GERL–Golgi; 2c, ER to vacuole transition; 2d, ER to protein body transition; 3a, Transport to, and deposition in, the vacuole (Vac.); 3b, Transport to plasmalemma and extracellular secretion; 4a, Vacuole to protein body/vacuole transition; 4b, Protein body to vacuole transition; 5, Endocytosis (Endocyt.) and associated intracellular pathways.

MVB, microvesicular bodies; PCR, partially coated reticulum, GERL, Golgi–ER–lysosome.

lysosome in animal cells is, however, clearly different. In contrast to the sorting mechanism in animal cells, yeast do not use mannose-6-phosphate receptors for vacuolar sorting, in fact glycosylation is not involved at all (Stevens *et al.* 1982). A similar situation is found in plants (Chrispeels 1985), where signal domains or patches on the surface of proteins are apparently used for sorting. Yeast may thus turn out to be a better model system for plant than for animal protein traffic.

The differential sorting and targeting of proteins within the plant endomembrane systems is still rather poorly understood in detail, but is known to depend upon characteristics of the molecules being transported, and various receptors and carriers that interact with them.

A series of questions may be posed:

1. What are the pathways of vesicle movement?
2. How is targeted transport achieved?
3. Does 'bulk flow' rule?
4. Is there a default pathway (in the absence of targeting information/instruction)?
5. Where do modifications to the vesicle contents occur?
6. What permits or prevents vesicle fusion?
7. Can we modify or change the pathways?

To answer such questions the following overview illustrates some of the advances that have, and are being, made.

Methods

A wide range of methods have been applied to the elucidation of intracellular transport pathways and their control. These include: fractionation of cellular components, with or without pulse-chase labelling; inhibitor studies, which have been particularly successful when applied with fractionation and/or cytochemical approaches; genetic studies, which have been used most effectively with yeast; and recently transgenic studies which have provided a new and particularly valuable technique with higher plants. Major advances in understanding intracellular transport in animal cells have resulted from *in vitro* reconstitution studies; these have so far only been followed to any extent using yeast components rather than those from higher plant cells.

The isolation, fractionation and analysis of specific subcellular components, and the extrapolation of their various interrelationships, is a well-established technique. Its application to studies of the plant endomembrane system has been extensively and excellently reviewed (see e.g. Robinson 1985). Organelle fractionation has proved to be particularly useful in association with pulse-chase labelling and inhibitor studies, although some care is required in interpretation of results. Faye & Chrispeels (1989), for example, have recently demonstrated that tunicamycin, an inhibitor of side-chain glycosylation and apparently secretion of β-fructosidase, acts not on the secretory process itself but by affecting the stability of the synthesised enzyme which, in its unglycosylated form, is degraded during or just after secretion.

Pulse-chase labelling and inhibitor studies have also been applied with microscopical analysis of intracellular transport pathways. The early electron microscopical (EM) autoradiography studies of Bailey *et al.* (1970) demonstrated the endoplasmic reticulum (ER) to protein body route for storage deposition in developing seed cotyledons. This developmental system has also been examined effectively by inhibitor studies (see e.g. Craig & Goodchild 1984) and by an immunocytochemical approach (for a review, see Herman 1988).

A wide range of other cytological techniques including the application of both optical and electron microscopy, have proved valuable in elucidating the organisation of the endomembrane system. For example, studies with both fluorescent (see e.g. Oparka *et al.* 1988; Cole *et al.* 1990) and electron opaque (see e.g. Tanchak *et al.* 1984; Joachim & Robinson 1984; Hillmer *et al.* 1986) markers have demonstrated the role of endocytosis in plant cells, and also the distribution of the labels taken up. Conventional (see e.g. Craig *et al.* 1980), confocal (see e.g. Cole *et al.* 1990), EM ultrathin (see e.g. Harris & Boulter 1976; Craig *et al.* 1979) and thick section/high voltage EM (see e.g. Marty 1978) studies have all been applied to the examination of vacuolar ontogeny.

The power of genetic analysis to investigate protein sorting has so far been applied only to yeast. A measure of the complexity of the secretory pathway can be seen in the large number (at least 40) of genes identified whose products are involved in the pathway. These *sec* mutations were selected as being more dense than wild-type cells and then screened for being defective for secretion at elevated temperature. They can be divided into several groups according to where in the secretory pathway they block transport (Novick *et al.* 1980).

A different strategy has been used to isolate mutants defective in transport from the Golgi to the vacuole. These *vps* mutations are unable to sort soluble vacuolar hydrolases such as carboxypeptidase Y (CPY) and proteinase A (PRA) correctly, and so secrete them via the default pathway. Selection strategies based on the secretion of either CPY or a CPY-invertase fusions have identified at least 40 complementation groups (Robinson *et al.* 1988; Rothman *et al.* 1989). There are probably many more *SEC* and *VPS* genes still to be identified.

Transgenic studies are proving to be particularly powerful in dissecting the function and control aspects of targeting in relation to intracellular transport in plants. In an extension of their earlier work on phaseolin, for example, Chrispeels' group demonstrated the correct targeting of the bean storage protein in seeds of transformed tobacco (Greenwood & Chrispeels 1985). Similarly Colot *et al.* (1987) were able to demonstrate transgenic expression dependent upon the endosperm-specific promoter of a cereal storage protein, again using an *Agrobacterium tumefaciens*/tobacco expression system. Such expression allows for the analysis of both promoter sequences, which control the temporal and spatial patterns of expression, and other sequence components that determine the characteristics which are responsible for the correct post-translational modifications and targeting. These analyses can be made using reporter genes which are relatively easy to assay; glucuronidase (GUS), for example, has proved to be particularly useful in histochemical studies of tissue-specific expression, although it was pointed out at the start (Jefferson *et al.* 1987) that particular care should be taken in interpretation of the results. More recently various laboratories have utilised anti-GUS for a finer, subcellular immunolocalisation of the reporter gene. The targeting requirements associated with the molecule itself are, however, also studied by use of constructs which result in modifications of, for example, the leader sequence, glycosylation sites, etc. of the protein. Transcriptional and translational fusions are utilised; in a transcriptional fusion the promoter of gene I is fused to the gene under study (II) so that the ATG translational start of gene II is active, whereas in a translational fusion the transcriptional and translational sites of the promoter gene (I) are utilised (the ATG of gene II may still be present, but not functional as a translational starting point).

The inclusion of presequences between a promoter, which is selected to give a high level of expression, and the reporter gene may influence the synthesis and targeting of the assayed product. Alternatively the use of a series in which small deletions are made from a functional construct allows analysis of those regions of the sequence which are responsible for particular aspects of expression and targeting of a protein.

Storage proteins have been examined extensively in transgenic studies, although in general these have required transformation and expression within the seed (usually tobacco) rather than in a transient assay system. There are practical difficulties in obtaining protoplasts for transient assay from starch-rich cotyledons/endosperm although, by contrast, protoplasts have been obtained from, for example, immature cotyledons of soybean (Lu *et al.* 1983).

It seems very likely that transgenic studies will continue to make an increasingly important contribution towards analysis of plant protein synthesis and intracellular transport/targeting and deposition/secretion. However, perhaps the only way in which a true molecular description of the events involved in protein sorting through the Golgi will be made is by a total *in vitro* reconstitution of vesicle traffic between organelles. The power of this technique has been demonstrated amply by the studies of Rothman and coworkers (see e.g. Balch *et al.* 1984), who have reconstituted vesicle traffic between the *cis* and *medial* compartments of mammalian Golgi. This *in vitro* assay has enabled them to fractionate the large array of cytosolic proteins essential for vesicular transport. Inhibitors such as *N*-ethylmaleimide (NEM) and GTP-γ-S have also helped to identify further proteins required for vesicle trafficking. Modifications to their system have allowed the events of vesicle budding and fusion to be studied.

Only recently have *in vitro* reconstitution assays been described for yeast. These rely on freeze-thawing of sphaeroplasts to produce ghosts, or gentle lysis of sphaeroplasts (Ruohola *et al.* 1988). Both of these techniques produce a heterogeneous mix of intact organelles lacking cytosolic proteins and small molecules. Transport is monitored by following the modification of *in vitro* synthesised ^{35}S-labelled pro-alpha mating factor from an ER- to a Golgi-modified form. Interestingly, neither of these systems reconstitute the later post-Golgi stages of secretion, for reasons that are not understood at present. As well as inhibitor and fractionation studies such as those pioneered by Rothman, these yeast systems are also open to genetic analysis. Cytosolic proteins prepared from *sec* strains are often unable to support transport at the elevated temperatures at which wild-type cytosol is still competent. Fractions of wild-type cytosolic proteins are used to complement this temperature sensitivity and thus identify the SEC gene product.

In vitro reconstitution of transport from the Golgi to the vacuole has recently been described. This uses the gentle lysis technique described above. However, the marker protein carboxypeptidase Y (CPY) is pulse-labelled in *in vitro* with ^{35}S. The pulse is 'frozen' at the Golgi stage by placing the cells on ice after a predetermined period. After gentle lysis the subsequent transport of CPY to the vacuole can be promoted by the addition of energy and cytosolic proteins. The mature vacuolar form of CPY is detected by immunoprecipitation and fluorography (Stevens 1990; Emr *et al.* 1990). Although at least 40 different genes involved in the transport of proteins from the Golgi to the vacuole have been described (Robinson *et al.* 1988; Rothman *et al.* 1989), none of them is temperature sensitive for sorting. Thus the *in vitro* complementation experiments described by Schekman for ER to Golgi reconstitution cannot be done for Golgi to vacuole sorting. The search is now on for mutations that are temperature sensitive for sorting to the vacuole.

Vacuole formation

The formation and function of plant cell vacuoles during cell enlargement and differentiation has been reviewed extensively (e.g. Marty *et al.* 1980; Boller & Wiemken 1986). The anastomosing cage of Golgi–ER–lysosome (GERL)-derived tubules described by Marty (1978) from his thick-section, high voltage EM studies of *Euphorbia* root tips also indicated the potential lysosomal function for the vacuole which was proposed by Matile earlier (e.g. 1975). In some cell types, however, it seems that ER dilation can give rise to the vacuole directly; this has been reported for both parenchyma cells and during storage deposition in endosperm (see e.g. Oparka & Harris 1982; Lending & Larkins 1989).

From ER to Golgi

The signal hypothesis (Blobel & Dobberstein 1975) for targeting of proteins to the ER and their subsequent transport to the ER lumen have been described in detail (for a review, see Walter & Lingappa 1986), although interestingly not all signal sequences appear to be functionally equivalent in a comparison, for example, between yeast and mammalian cells (Bird *et al.* 1987).

By analogy with vesicle transport in mammalian Golgi, we can expect that ER to Golgi vesicle transport is likely to be equally as complex. There is also the added complication that there is two-way vesicle traffic between the Golgi and ER. It is now generally accepted that lumenal ER proteins are retained in the ER by a recycling mechanism that involves receptors in the *cis*-Golgi and a C-terminal signal domain of HDEL on all the ER lumenal proteins (Pelham, 1988; Pelham *et al.* 1988).

Ten yeast SEC genes involved in transport from the ER to the Golgi have been identified (Novick *et al.* 1980). At the restrictive temperature of 37 °C mutations in these genes cause the accumulation of exaggerated ER structures, some also accumulate numerous vesicles. Those mutations that cause only the accumulation of ER are presumed to be blocked prior to vesicle budding, whereas those that cause vesicle accumulation are blocked before vesicle fusion with the *cis*-Golgi. Eight of these mutations have been studied in detail. A clue as to the interactions of the products of these genes can be seen in studies with double-mutant strains (Schekman *et al.* 1990). Some pairs of mutations exhibit very poor growth even at the permissive temperature where growth of single mutations is normal. Of five mutations causing defects prior to vesicle budding, *sec13* and *sec16* interact with each other, and *sec12* and *sec23* interact with both of these. However, *sec21* interacts with none of these. Of three mutations blocked after vesicle budding, *sec17* and *sec18* interact, but neither does so with *sec22*.

Many of these *SEC* genes have been cloned by complementation of the temperature-sensitive defect. The nature of their protein products has been deduced from sequence analysis. Due to the *in vitro* reconstitution assay described above for ER to Golgi transport, an initial functional characterisation of some of these proteins has also been achieved. The *SEC23* gene product (Sec23p) is a predicted 85 kDa protein which, on the basis of *in vitro* complementation assays, has now been purified 10 000-fold. The complementing activity consists of two polypeptides, one of 85 kDa (Sec23p) and one of 105 kDa (possibly the product of *sec21*), both are peripheral membrane-associated proteins on the cytoplasmic side (Nakano *et al.* 1988).

The partial purification of Sec13p has also been accomplished using an *in vitro* complementation assay. It is a 34 kDa cytosolic protein that also forms a complex, distinct from that of Sec23p. The *SEC12* gene codes for a 52 kDa protein that forms a 70 kDa glycoprotein *in vivo*. It has both N- and O-linked glycosylation and is an integral membrane protein found in both the ER and the Golgi (Nakano *et al.* 1988). It has an active cytoplasmic domain that interacts with Sec13p, Sec16p and the product of the *GAR1* gene. However, the role of all of these proteins in vesicle budding from the ER is unknown.

Further insight into the precise mechanism by which these *sec* gene products promote the fusion of ER-derived vesicles with the *cis*-Golgi has come from the work of Rothman on mammalian Golgi. He has identified a protein that is sensitive to NEM (called NSF for NEM-sensitive factor) that promotes fusion of Golgi transport vesicles with recipient cisternae (Malhotra *et al.* 1988). In yeast the product of the *SEC18* gene is homologous to mammalian NSF (Balch *et al.* 1984). Thus NSF is a highly conserved protein required for vesicle fusion at many stages through the secretory pathway. NSF itself is not a 'fusogen' and requires other proteins to allow NSF to bind to membranes. These are termed SNAP proteins (soluble NSF attachment proteins). Three SNAP proteins have been identified (α, β and α), of which α-SNAP can restore activity to a *sec17*-derived cytosol in the yeast *in vitro* reconstitution assay (Rothman 1990). This implies that *SEC17* is the structural gene for yeast α-SNAP, and confirms at a biochemical level the genetic observation of interaction between the products of *SEC17* and *SEC18*.

The degree of functional conservation between mammalian and yeast systems suggests that it is unlikely that plants will be any different. The basic mechanisms of vesicle transport and the proteins that are required for this will probably be shown to be the same or similar in all eukaryotic organisms. Major differences between plants and animals may, however, be found in the receptor protein molecules by which vesicles from one compartment recognise their target membrane. The nature of these receptors is at present unknown for any system.

From Golgi to vacuole

Plant proteins destined for either internal deposition within the vacuole, or external secretion, are usually transported, and often modified post-translationally, by the Golgi apparatus (for reviews, see Chrispeels 1985; Harris 1986). There is, however, generally no evident, direct connection between the ER and the Golgi in cells which are involved primarily in biosynthetic processes (Fig. 2 (a)). In cells in which controlled biodegradation is a major process, however, connection between the ER and Golgi, via the GERL (Fig. 2 (b)) has been reported (see e.g. Marty 1978; Harris & Oparka 1983). This difference in ER/Golgi structural interrelationship at different stages of the cell cycle is exemplified by storage parenchyma cells in seed cotyledons. During seed development there is synthesis and deposition of storage proteins (Fig. 2 (a)) but then, in a carefully regulated process (Chrispeels *et al.* 1976), the storage proteins are mobilised by *de novo* synthesised endopeptidases which are transported to, and function within, the protein bodies/vacuoles (Fig. 2 (b)).

Cells are, however, not involved in just one biosynthetic or degradative function at a time. Whilst the pathways associated with storage protein biosynthesis and deposition may be a major activity, simultaneous carbohydrate and protein secretion associated with cell wall synthesis is occurring but by a pathway which is divergent from the Golgi. By contrast, the endocytotic and lysosomal pathways may be convergent (Record & Griffing 1988).

How is sorting to the various destinations (Fig. 1) achieved? Evidence from yeast and plant studies has identified components of both the transporting vesicles and their contents which are involved in this sorting.

The yeast Golgi apparatus is a poorly defined organelle usually only visible by EM in aberrant strains. Few mutations have been described that are defective in just Golgi vesicle transport. *SEC14* encodes a 35 kDa cytosolic protein required for exit of protein traffic from the Golgi (Bankaitis *et al.* 1989). Yeast *sec14* strains accumulate Golgi modified forms of both secretory and vacuolar proteins and also presumed Golgi-derived vesicles at the restrictive temperature. Due to the lack of a suitable *in vitro* assay, the biochemical role of Sec14p has not been elucidated further.

Yeast *sec7* strains exhibit grossly exaggerated Golgi structures and accumulate glycoproteins at various stages of outer chain modification when grown at the restrictive temperature. Sec7p is a large 230 kDa cytosolic protein required for both exit from the Golgi and intra-Golgi transport (Franzusoff & Schekman 1989).

Another gene identified as being involved in Golgi vesicle transport is

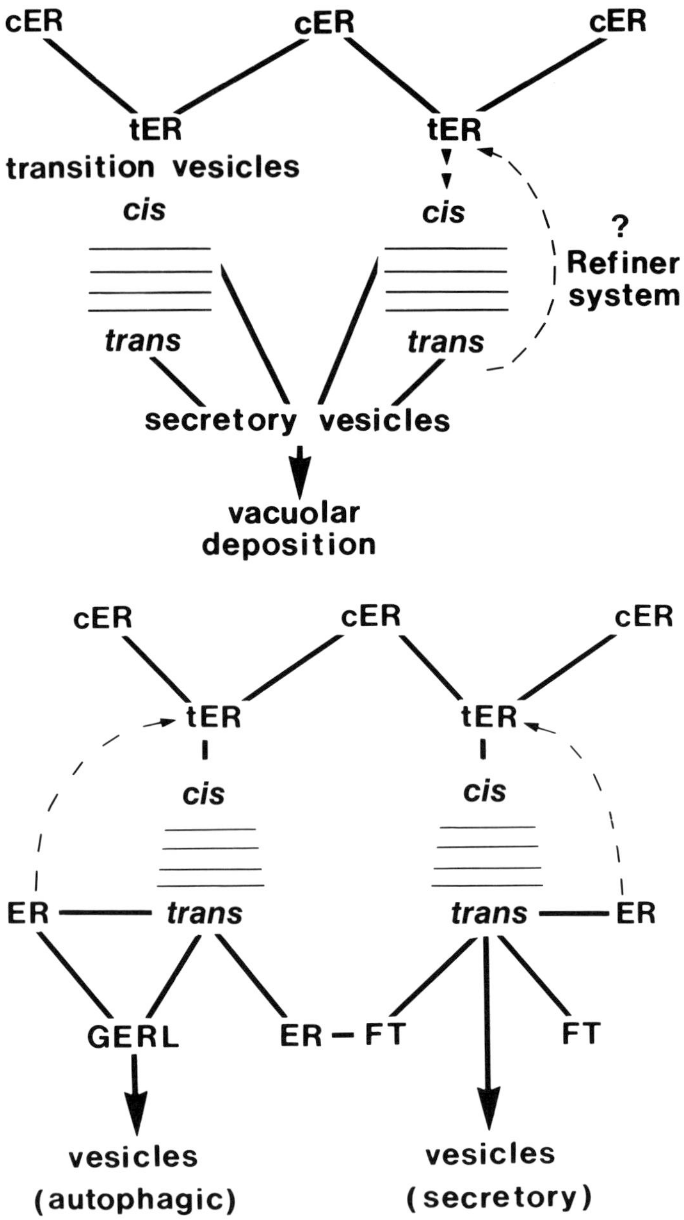

Fig. 2. Structural interrelationships of endomembrane system components in seed cotyledonary storage parenchyma cells during (a) storage protein synthesis and deposition and (b) storage protein mobilisation. CER, cisternal ER; tER, tubular ER; GERL, Golgi–ER–lysosome; FT, fine tubules.

YPT1. The product of this gene is a *ras*-like GTP-binding protein. Yeast *ypt1* strains accumulate both secretory and vacuolar proteins in Golgi modified forms (Bacon *et al.* 1989). GTP-binding proteins are involved in many stages of the secretory pathway.

The pathway of vesicle transport from the Golgi, to the vacuole is complex. The 40 *VPS* genes identified as being important in this process can be divided into three phenotypic classes. Class A has normal vacuoles, class B many small vacuoles, and class C no visible vacuoles but often accumulates small vesicles. One could postulate that these represent three stages of a pathway Golgi–class C–class B–class A–vacuole. Not all of the class A mutations may be directly involved in this pathway, but may be defects in the cytoskeleton or vacuole acidification. Although, contrary to recently held theories, vacuole acidification does not seem necessary for the correct sorting of vacuolar proteins (Stevens 1990). Some of the class B and C mutations are sensitive to growth on 1.5 M NaCl; this has enabled further mutations to be isolated, many of which are severely defective in vacuole biogenesis. At least four to five new genes have been identified by this selection process (M. D. Watson & M. Latterich, unpublished observations).

None of the *VPS* genes has been characterised via the *in vitro* reconstitution assay. However, nine of them have been cloned and five sequenced. The products of *VPS15*, *-16*, *-18*, *-33* and *-34* are all large proteins which associate with membranes on the cytoplasmic side. Vps33p is able to bind ATP, is not glycosylated and is a rare protein. Vps15p is a large protein of approximately M_r 145 000 that has homology with serine-threonine protein kinases. It is also anchored to membranes by myristylate at the N-terminal end. A speculative role for this protein is that it is anchored to vesicles rather than to the vacuolar membrane (Emr *et al.* 1990).

Studies of targeting to the vacuole of higher plants have involved biochemical analyses, inhibitor studies and transgenic expression in both plants (usually tobacco) and yeast. By analogy with the more widely studied transport systems in animal cells, where protein glycosylation is known to play an important role in targeting (for a review, see Paulson 1989), early work on plant proteins concentrated on analysis of glycosylation (for a review, see Chrispeels *et al.* 1987). However the facts that (1) not all proteins which are deposited in the vacuole are glycoproteins and (2) tunicamycin, an inhibitor of glycosylation, did not affect phytohaemagglutinin (PHA) targeting to the vacuole (Bollini *et al.* 1985), indicated that either glycosylation was not essential or that this route represents a default pathway, i.e. the pathway taken when there is no specific targeting information. The default pathway in both animal and yeast cells is, however, external secretion.

Several studies (e.g. Bowles *et al.* 1986; Craig & Goodchild 1984) have utilised monensin to perturb the normal pathway of storage protein deposition from Golgi to vacuole/protein body. The ionophore induces ion imbalance in the endomembrane system, although it is not yet clear whether the misdirection of vesicles with storage protein contents is a consequence of disruption of (1) sorting at the Golgi, (2) targeting or (3) recognition events associated with membrane fusion. However, the consequences of monensin treatment are that proteins normally deposited into the vacuole are secreted to the paramural space, suggesting that this route might be the default pathway in plant cells as in animal and yeast. Denecke *et al.* (1990) provided more direct evidence that this default pathway could exist in plants; they established an *in vivo* system for transient expression of constructs which included ER-targeting sequences attached to cytosolic enzymes from prokaryotes which were used as reporters. Irrespective of the ER signal peptide used, each of the enzymes was secreted from the protoplasts, suggesting that such secretion is the default pathway (see Jones *et al.*, this volume).

If the default pathway is external secretion and glycosylation is not required for targeting to the vacuole, how are storage proteins routed to their vacuolar destination? Extensive and careful analyses by Chrispeels and co-workers have shown that specific domains on storage proteins are associated with vacuolar targeting (Tague *et al.* 1990), and that transported proteins, even when there are high rates of bulk flow through the endomembrane system, still require positive sorting information for targeting to the vacuole (Dorel *et al.* 1989).

Having established that the plant storage protein PHA could be expressed and transported correctly in yeast under the control of a yeast acid phosphatase promoter (Tague & Chrispeels 1987), a series of 3′ deletions of PHA was made with in-frame fusion to invertase, which was utilised as a reporter (Tague *et al.* 1990). A targeting domain was identified in the PHA sequence that was similar to the yeast CPY sorting domain. An interesting additional result from this work was the observation that when a single amino acid alteration to the sequence resulted in glycosylation close to the centre of the sorting domain the protein was secreted, i.e. the targeting sequence was 'covered' and the protein then reverted to the default pathway.

But do such studies involving expression of plant genes in yeast tell us about targeting of plant proteins or just about yeast transport mechanisms? Certainly the experimental system is attractive and has been used successfully. However, in a comparative study in which a storage protein cDNA was expressed correctly in homologous and heterologous plant (tobacco) systems only limited processing was observed in the yeast system (Croy *et al.* 1988). A full-length cDNA, from the *11S* legumin gene family, including coding for an authentic leader sequence was placed downstream of yeast *PGK* (the gene for phosphoglycerokinase) and yielded high levels of immuno-detectable protein in yeast. The protein synthesised was, however, an unreduced polypeptide of M_r 60 000 unlike the mature legumin, which is proteolytically processed to disulphide-linked polypeptides with M_r values of 38 000 and 20 000. Immunolabelling showed that the protein was restricted to the ER and vesicles associated with the Golgi; one interpretation suggests that the plant leader sequence was not sufficiently well recognised in the yeast system and was not cleaved (Croy *et al.* 1988). In the same work the legumin gene was expressed and processed in tobacco as in pea seed development, although with the interesting addition that the protein was synthesised and deposited in the tobacco endosperm as well as in the embryo. In pea the endosperm is transient, being crushed by the enlarging embryo, although both legumin mRNA and protein accumulate in the endosperm prior to embryo expansion (Harris *et al.* 1989).

Endosperm and embryo storage reserves are mobilised during germination in quite different patterns. Whilst storage proteins in embryo tissues are degraded within the vacuolar apparatus, with a reversion of structural form from numerous small protein storage vacuoles to one or a few large 'empty' vacuoles per cell (Fig. 1), storage mobilisation in cereal starchy endosperm is regulated not within the tissue, which is dead, but by the rate of release of enzymes from aleurone and embryo. Are the cereal endosperm storage proteins synthesised and deposited as in dicotyledon embyros or by different pathways? Synthesis of both aqueous-soluble albumin proteins and salt-soluble globulins occurs in cereal endosperm and dicotyledon embryos, but the cereals also produce prolamins, which are soluble in aqueous alcohols, and glutenins, which are extracted after the other proteins by alkaline or acidic solvents. The prolamins are accumulated within membrane-bounded protein bodies but the ontogeny of these structures is controversial. In maize, protein bodies form by dilation of the ER (Khoo & Wolf 1970; Larkins & Hurkman 1978; Lending & Larkins 1989) (Figs. 1 and 2 (d)), but opinions vary as to what occurs in barley and wheat. Cameron-Mills & von Wettstein (1980), Parker (1980) and Bechtel *et al.* (1982) suggested that the prolamins are deposited in a manner similar to that seen for embryo globulins, whilst Miflin *et al.* (1983) suggested that the important parallel is with maize. Cameron-Mills & von Wettstein (1980) were, however, unable to find any continuity between ER and protein bodies in barley comparable with that seen in maize (Khoo & Wolf 1970; Lending & Larkins 1989). More recently Henderson (1987) has suggested a model for prolamin synthesis and deposition in barley which accounts for new results and those used to support the previous and apparently contradicting models.

Wilkins *et al.* (1990) have reported the expression of barley lectin, a soluble vacuolar protein which is glycosylated only in its proprotein form, in leaf vacuoles in transgenic tobacco. Whilst lectin from both wild-type (wt) and a glycosylation-minus (gly^-) mutant were expressed under transcriptional control of the CaMV35S promoter, the gly^- proprotein was processed more rapidly to the mature protein and also transported from the Golgi faster than the wt proprotein. As with jack bean pro-ConA (see Bowles & Pappin 1988, and references therein) and pea pro-vicilin (Craig & Goodchild 1984) monensin was inhibitory to processing of the barley prolectin; the implication is that deglycosylation may occur, perhaps as a rate-limiting step in the *trans*-Golgi.

What about transport and post-translational modification within the Golgi? The animal cell Golgi has been examined extensively by Rothman and co-workers (see e.g. Pfeffer & Rothman 1987; Beckers *et al.* 1989; Diaz *et al.* 1989; see also Hawes *et al.*, this volume). Less detail is known about the functioning of the plant Golgi apparatus, although its central role in intracellular transport and modification of proteins and carbohydrates is well established (see e.g. Chrispeels 1984, 1985; Grief & Shaw 1987; Staehlin, this volume). The manner in which vesicles are budded and fused is uncertain, and we have only preliminary details about the relationship with other structural features such as the partially coated reticulum (PCR) (Tanchak *et al.* 1988; Hillmer *et al.* 1988). Robinson's group (see e.g. Robinson *et al.*, this volume) and Harley & Beevers (1989) have identified pea cotyledons as a rich source of plant clathrin-coated vesicles. These vesicles have associated storage protein and a role in transport would seem likely, presumably from the Golgi: interestingly such a vast population of coated vesicles has not been reported in any of the ultrastructural studies of developing pea cotyledons, although they have been studied extensively using both osmicated and non-osmicated tissues (see Craig *et al.* 1979, 1980).

The individual dictyosomes of plant cells are less clearly interrelated than those in animal cells, although serial and thick-sectioning has revealed the presence of interconnecting fine tubules (see e.g. Harris & Oparka 1983), the GERL (see e.g. Marty 1978) and the *trans*-Golgi network. The full range of endomembrane interrelationships remains to be determined, although detailed and quantitative studies of some systems (see e.g. Steer 1988; Steer & Steer 1989; Steer, this volume) have been undertaken.

Conclusions

It is now clear that in many plant tissues secretion of proteins is a default pathway as it is in animals and yeast (see e.g. Denecke *et al.* 1990; Holwerda *et al.* 1990). Sorting to the vacuole requires positive sorting signals on the surface of vacuolar proteins. The nature of these sorting signals remains to be determined, but it is known that glycan side-chains are not essential, but may influence the rate of post-translational processing and transport. Preliminary evidence for the sorting of barley lectin to the vacuole shows that the sorting information resides in a C-terminal propeptide (Bednarek *et al.* 1990), and Tague *et al.* (1990) have identified a domain on a plant protein which targets a marker gene to the yeast vacuole.

As in plants, yeast vacuolar proteins are not dependent upon glycosylation for sorting but have proteinaceous signal domains (Stevens *et al.* 1982). The enzyme CPY is taken as a model soluble vacuolar protein: an N-terminal propeptide of 111 amino acid residues contains the four residue motif of QRPL near its N-terminal end. This motif is both necessary and sufficient for the sorting of CPY to the vacuole (Valls *et al.* 1987). Proteinase A also has a propeptide necessary for vacuolar targeting; however, it bears little homology to that of CPY (Klionsky *et al.* 1988). Vacuolar integral membrane proteins are probably sorted to the vacuole by a different mechanism. Dipeptidyl aminopeptidase B and alkaline phosphatase are both anchored to the membrane via a single transmembrane domain that acts as a signal-anchor sequence. The bulk of the proteins is found in the lumen of the vacuole with just a short cytoplasmic tail remaining on the N-terminal end. Sorting of both these enzymes seems to require only the transmembrane domain (Klionsky & Emr 1989; Roberts & Stevens 1990). Such a structure is, however, quite different from that recently reported for the plant tonoplast intrinsic protein (Johnson *et al.* 1990).

A series of questions was posed at the start of this chapter. How can we begin to answer them?

1. Some of the pathways of vesicle movement have been followed, but there are many points which remain unclear. For example, how do the transition vesicles, which travel from ER to Golgi, form? What is the nature and function of the non-clathrin coating which seems to be associated with intra-dictyosome vesicle movement? What is the relationship of the PCR to other parts of the endomembrane system? etc.
2. As to how targeted transport is achieved, we know that vesicle

contents are important, although with regard to proteins it appears that the critical features lie in domains of the tertiary structure rather than glycan side-chain components.
3. Bulk flow does not rule; a vacuolar targeting sequence on a specific protein would still seem to be required even when the majority of synthesised protein is being directed to the vacuole (Dorel *et al.* 1989).
4. The default pathway for material which enters the ER lumen in plants would appear to be extracellular secretion, as in animal and yeast cells.
5. Post-translational modifications can occur within the ER, Golgi and vacuole, and their extent and rate may be modulated by glycan side-chains.
6. Animal and yeast studies have described both membrane and cytosolic factors which are required for vesicle fusion.
7. Transgenic studies have shown that it is possible to modify or change the normal pathway by minor modifications to the component being transported.

Clearly we are starting to answer some of the questions posed above but in doing so even more questions seem to be raised.

References

Bacon, R.A., Salminen, A., Ruohola, H., Novick, P. & Ferro-Novic, S. (1989). The GTP-binding protein YptI is required for transport *in vitro*: the Golgi apparatus is defective in YptI mutants. *Journal of Cell Biology*, **109**, 1015–22.

Bailey, C.J., Cobb, A. & Boulter, D. (1970). A cotyledon slice system for the electron autoradiographic study of the synthesis and intracellular transport of seed storage proteins of *Vicia faba*. *Planta*, **95**, 103–8.

Balch, W.E., Dunphy, W.G., Braell, W.A. & Rothman, J.D. (1984). Reconstitution of the transport of protein between successive compartments of the Golgi measured by the coupled incorporation of N-acetylglucosamine. *Cell*, **39**, 405–16.

Bankaitis, V.A., Melehorn, D.E., Emr, S.D. & Green, R. (1989). The *Saccharomyces cerevisiae SEC14* gene encodes a cytosolic factor that is required for transport of secretory proteins from the yeast Golgi complex. *Journal of Cell Biology*, **108**, 1271–81.

Bechtel, D.B., Gaines, R.L. & Pomeranz, Y. (1982). Early stages in wheat formation and protein body initiation. *Annals of Botany*, **50**, 507–18.

Beckers, C.J.M., Block, M.R., Glick, B.S., Rothman, J.E. & Beach, W.E. (1989). Vesicular transport between the endoplasmic reticulum and the Golgi stack requires the NEM-sensitive fusion protein. *Nature (London)*, **339**, 397–8.

Bednarek, S.Y., Wilikins, T.A. & Raikhel, N.V. (1990). Sorting mechanisms of barley lectin to the vacuoles of plant cells. *Journal of Cellular Biochemistry*, Suppl. **14C** (4312), p. 55.

Bird, P., Gething, M-J. & Sambrook, J. (1987). Translocation in yeast and mammalian cells: not all signal sequences are functionally equivalent. *Journal of Cell Biology*, **105**, 2905–14.

Blobel, G. & Dobberstein, B. (1975). Transfer of protein across cell membranes. *Journal of Cell Biology*, **67**, 835–51.

Boller, T. & Wiemken, A. (1986). Dynamics of vacuolar compartmentation. *Annual Review of Plant Physiology*, **37**, 137–64.

Bollini, R., Ceriotti, A., Daminati, M.G. & Vitale, A. (1985). Glycosylation is not needed for the intracellular transport of phytohaemaglutinin in developing *Phaseolus vulgaris* cotyledons and for the maintenance of its biological activities. *Physiologia Plantarum*, **65**, 15–22.

Bowles, D.J., Marcus, S.E., Pappin, D.J.C., Findlay, J.B.C., Maycox, P.R. & Burgess, J. (1986). Post-translational processing of concanavalin A precursors in jackbean cotyledons. *Journal of Cell Biology*, **102**, 1284–97.

Bowles, D.J. & Pappin, D.J. (1988). Traffic and assembly of concanavalin A. *Trends in Biochemical Science*, **13**, 60–4.

Cameron-Mills, V. & von Wettstein, D. (1980). Protein body formation in the developing barley endosperm. *Carlsberg Research Communications*, **45**, 577–94.

Chrispeels, M.J. (1984). Biosynthesis, processing and transport of storage proteins and lectins in cotyledon and developing legume seeds. *Philosophical Transactions of the Royal Society London Ser. B. Biological Sciences*, **304**, 309–22.

Chrispeels, M.J. (1985). The role of the Golgi apparatus in the transport and post-translational modification of vacuolar (protein body) proteins. *Oxford Surveys of Plant Molecular and Cell Biology*, **2**, 43–68.

Chrispeels, M.J., Baumgartner, B. & Harris, N. (1976). Regulation of reserve protein metabolism in the cotyledons of mung bean seedlings. *Proceedings of the National Academy of Sciences, USA*, **73**, 222–45.

Chrispeels, M.J., Faye, L., Sturm, A. & Johnson, K.D. (1987). Are oligosaccharide sidechains necessary for glycoprotein transport in targeting? In *Plant membranes: structure, function, biogenesis*, ed. C.J. Leaver & H. Sze, pp. 275–87. New York: Alan R. Liss.

Cole, L., Coleman, J., Evans, D. & Hawes, C. (1990). Internalisation of FITC and FITC-dextran by suspension-cultured plant cells. *Journal of Cell Science* (in press).

Colot, V., Robert, L.S., Kavanagh, T.A., Bevan, M.W. & Thompson, R.D. (1987). Localization of sequences in wheat endosperm protein genes which confer tissue-specific expression in tobacco. *EMBO Journal*, **6**, 3559–64.

Craig, S. & Goodchild, D.J. (1984). Golgi-mediated vicilin accumulation in pea cotyledon cells is redirected by monensin and nigericin. *Protoplasma*, **122**, 91–7.

Craig, S., Goodchild, D.J. & Hardham, A.R. (1979). Structural aspects of protein accumulation in developing pea cotyledons. I. Qualitative and quantitative changes in parenchyma cell vacuoles. *Australian Journal of Plant Physiology*, **6**, 81–98.

Craig, S., Goodchild, D.J. & Miller, C. (1980). Structural aspects of protein accumulation in developing pea cotyledons. II. Three-dimensional reconstruction of vacuoles and protein bodies from serial sections. *Australian Journal of Plant Physiology*, **7**, 329–37.

Croy, R.R.D., Evans, I.M., Yarwood, J.N., Harris, N., Gatehouse, J.A., Shirsat, A.H., Kang, A., Ellis, J.R., Thompson, A. & Boulter, D. (1988). Expression of pea legumin sequences in pea, *Nicotiana* and yeast. *Biochimie und Physiologie der Pflanzen*, **183**, 183–97.

Denecke, J., Botterman, J. & Deblaere, R. (1990). Protein secretion in plant cells can occur via a default pathway. *Plant Cell*, **2**, 51–9.

Diaz, R., Mayorga, L.S., Weidman, P.J., Rothman, J.E. & Stahl, P.D. (1989). Vesicle fusion following receptor-mediated endocytosis requires a protein active in Golgi transport. *Nature (London)*, **339**, 398–400.

Dorel, C., Voelker, T.A., Herman, E.M. & Chrispeels, M.J. (1989). Transport of the proteins to the plant vacuole is not by bulk flow through the secretory system, and requires positive sorting information. *Journal of Cell Biology*, **108**, 327–38.

Emr, S., Herman, P., Banta, L., Robinson, J., Klionsky, D. & Vida, T. (1990). Vacuolar protein sorting in yeast: potential role for a protein kinase. *Journal of Cellular Biochemistry*, Suppl. 14C, H009, p. 6.

Faye, L. & Chrispeels, M.J. (1989). Apparent inhibition of β-fructosidase secretion by tunicamycin may be explained by breakdown of the unglycosylated protein during secretion. *Plant Physiology*, **89**, 845–51.

Franzusoff, A. & Schekman, R. (1989). Functional compartments of the yeast Golgi apparatus are defined by the *sec7* mutation. *EMBO Journal*, **8**, 2695–702.

Greenwood, J.S. & Chrispeels, M.J. (1985). Correct targeting of the bean storage protein phaseolin in the seeds of transformed tobacco. *Plant Physiology*, **79**, 65–71.

Grief, C. & Shaw, P.J. (1987). Assembly of cell-wall glycoprotein of *Chlamydomonas reinhardii*: oligosaccharides are added in medial and trans Golgi compartments. *Planta*, **171**, 302–12.

Harley, S.M. & Beevers, L. (1989). Coated vesicles are involved in the transport of storage proteins during seed development in *Pisum sativum* L. *Plant Physiology*, **91**, 674–8.

Harris, N. (1986). Organization of the endomembrane system. *Annual Review of Plant Physiology*, **37**, 73–92.

Harris, N. & Boulter, D. (1976). Protein body formation in cotyledons of developing cowpea (*Vigna unguiculata*) seeds. *Annals of Botany*, **40**, 739–44.

Harris, N., Grindley, H., Mulcrone, J. & Croy, R.R.D. (1989). Correlated *in situ* hybridisation and immunocytochemical studies of legumin storage protein deposition in pea (*Pisum sativum* L.). *Cell Biology International Reports*, **13**, 23–35.

Harris, N. & Oparka, K.J. (1983). Connections between dictyosomes, ER and GERL in cotyledons of mungbean (*Vigna radiata* L.). *Protoplasma*, **114**, 93–102.

Henderson, J. (1987). Protein deposition in developing barley endosperm. Ph.D. thesis, University of Durham.

Herman, E.M. (1988). Immunocytochemical localization of macro-molecules with the electron microscope. *Annual Review of Plant Physiology and Plant Molecular Biology*, **39**, 139–55.

Hillmer, S., Depta, H. & Robinson, D.G. (1986). Confirmation of endocytosis in higher plant protoplasts using lectin–gold conjugates. *European Journal of Cell Biology*, **41**, 142–9.

Hillmer, S., Freundt, H. & Robinson, D.G. (1988). The partially coated reticulum and its relationship to the Golgi apparatus in higher plant cells. *European Journal of Cell Biology*, **47**, 206–12.

Holwerda, B.C., Galvin, N.J. & Rogers, J.C. (1990). Processing and targeting of aleurain, a cysteine protease in barley aleurone cells. *Journal of Cellular Biochemistry*, Suppl. 14, H209, p. 40.

Jefferson, R.A., Kavanagh, T.A. & Bevan, M.W. (1987). GUS fusions: glucuronidase as a sensitive and versatile gene-fusion marker in higher plants. *EMBO Journal*, **6**, 3901–7.

Joachim, S. & Robinson, D.G. (1984). Endocytosis of cationic ferritin by bean leaf protoplasts. *European Journal of Cell Biology*, **34**, 212–16.

Johnson, K.D., Hofte, H. & Chrispeels, M.J. (1990). An intrinsic tonoplast protein of protein storage vacuoles in seeds is structurally related to a bacterial solute transporter (GIpF). *Plant Cell*, **2**, 525–32.

Khoo, U. & Wolf, M.J. (1970). Origin and development of protein granules in maize endosperm. *American Journal of Botany*, **57**, 1042–50.

Klionsky, D.J., Banta, L.M. & Emr, S.D. (1988). Intracellular sorting and processing of a yeast vacuolar hydrolase: proteinase A propeptide contains vacuolar targeting information. *Molecular and Cellular Biology*, **8**, 2105–16.

Klionsky, D.J. & Emr, S.D. (1989). Membrane protein sorting: biosynthesis, transport and processing of yeast vacuolar alkaline phosphatase. *EMBO Journal*, **8**, 2241–50.

Larkins, B.A. & Hurkman, W.J. (1978). Synthesis and deposition of zein in protein bodies of maize endosperm. *Plant Physiology*, **62**, 256–63.

Lending, C. R. & Larkins, B. A. (1989). Changes in the zein composition of protein bodies during maize endosperm development. *Plant Cell*, **1**, 1001–23.

Lu, D. Y., Cooper-Brand, S., Pental, D., Cocking, E. G. & Davey, M. R. (1983). Isolation and sustained division of protoplasts from cotyledons of seedlings and immature seeds of *Glycine max* L. *Zeitschrift für Pflanzenphysiologie*, **111**, 389–94.

Malhotra, V., Orci, L., Glick, B. S., Block, M. R. & Rothman, J. E. (1988). Role of an *N*-ethylmaleimide-sensitive transport component in promoting fusion of transport vesicles with cisternae of the Golgi stack. *Cell*, **54**, 221–7.

Marty, F. (1978). Cytochemical studies on GERL, provacuoles and vacuoles in root meristematic cells of *Euphorbia*. *Proceedings of the National Academy of Sciences, USA*, **75**, 852–6.

Marty, F., Branton, D. & Leigh, R. A. (1980). Plant vacuoles. In *The Biochemistry of Plants*, ed. N. E. Tolbert, pp. 625–59. New York: Academic Press.

Matile, P. (1975). *The lytic compartment of plant cells*. Springer-Verlag, Berlin & New York.

Miflin, B. J., Field, J. M. & Shewry, P. R. (1983). Cereal storage proteins. In *Seed Proteins*, ed. J. Mosse, J. Danssant & J. Vaughan, pp. 255–319. New York: Academic Press.

Nakano, A., Brada, D. & Schekman, R. (1988). A membrane glycoprotein from the endoplasmic reticulum to the Golgi apparatus in yeast. *Journal of Cell Biology*, **107**, 851–63.

Novick, P., Field, C. & Schekman, R. (1980). Identification of 23 complementation groups required for post-translational events in the yeast secretory pathway. *Cell*, **21**, 205–15.

Oparka, K. J. & Harris, N. (1982). Rice protein body formation: all types are initiated by dilation of the ER. *Planta*, **154**, 184–8.

Oparka, K. J., Robinson, D., Prior, D. A. M., Derrick, P. & Wright, K. M. (1988). Uptake of Lucifer Yellow CH into intact barley roots: evidence for fluid-phase endocytosis. *Planta*, **176**, 541–7.

Parker, M. L. (1980). Protein body inclusions in developing wheat endosperm. *Annals of Botany*, **46**, 29–36.

Paulson, J. C. (1989). Glycoproteins: what are the sugar chains for? *Trends in Biochemical Science*, **14**, 272–5.

Pelham, H. R. B. (1988). Evidence that luminal ER proteins are sorted from secreted proteins in a part ER compartment. *EMBO Journal*, **7**, 913–18.

Pelham, H. R. B., Hardwick, K. G. & Lewis, M. J. (1988). Sorting of soluble ER proteins in yeast. *EMBO Journal*, **7**, 1757–62.

Pfeffer, S.R. & Rothman, J.E. (1987). Biosynthetic protein transport and sorting by the endoplasmic reticulum and Golgi. *Annual Review of Biochemistry*, **56**, 829–52.

Record, R.D. & Griffing, L.R. (1988). Convergence of the endocytotic and lysosomal pathways in soybean protoplasts. *Planta*, **176**, 425–32.

Roberts, C. & Stevens, T.H. (1990). Localization of the yeast vacuolar membrane protein DRAP B. *Journal of Cellular Biochemistry*, Suppl. 14, H308, p. 54.

Robinson, D.G. (1985). *Plant Membranes: Endo- and Plasma Membranes of Plant Cells*. New York: John Wiley & Sons.

Robinson, J.S., Klionsky, D.J., Banta, L.M. & Emr, S.D. (1988). Protein sorting in yeast: isolation of mutants defective in the delivery and processing of multiple vacuolar hydrolases. *Molecular and Cellular Biology*, **8**, 4936–48.

Rothman, J.E. (1990). A molecular dissection of vesicular transport. *Journal of Cellular Biochemistry*, Suppl. 14C, H024, p. 13.

Rothman, J.H., Howard, I. & Stevens, T.H. (1989). Characterization of genes required for protein sorting and vacuolar function in the yeast *Saccharomyces cerevisiae*. *EMBO Journal*, **8**, 2057–65.

Ruohola, H., Kabcenell, A.K. & Ferro-Novick, S. (1988). Reconstitution of protein transport from the endoplasmic reticulum to the Golgi complex in yeast: the acceptor Golgi compartment is defective in the *sec23* mutant. *Journal of Cell Biology*, **107**, 1465–76.

Schekman, R., Baker, D., D'Enfert, C., Hicke, L., Hosobuchi, M., Kaiser, C., Pryer, N. & Rexach, M. (1990). Molecular requirements for vesicle budding and fusion in the secretory pathway. *Journal of Cellular Biochemistry*, Suppl. 14C, H025, p. 14.

Steer, M.W. (1988). Plasma membrane turnover in plant cells. *Journal of Experimental Botany*, **39**, 987–96.

Steer, M.W. & Steer, J.M. (1989). Pollen tube tip growth. *New Phytologist*, **111**, 323–58.

Stevens, T.H. (1990). Protein targeting to the yeast vacuoles. *Journal of Cellular Biochemistry*, Suppl. 14C, H011, p. 6.

Stevens, T.H., Esmon, B. & Schekman, R. (1982). Early stages in the yeast secretory pathway are required for transport of carboxypeptidase Y to the vacuole. *Cell*, **30**, 439–48.

Tague, B.W. & Chrispeels, M.J. (1987). The plant vacuolar protein, phytohemagglutinin, is transported to the vacuoles of transgenic yeast. *Journal of Cell Biology*, **105**, 1971–9.

Tague, B.W., Dickinson, C.D. & Chrispeels, M.J. (1990). A short domain of the plant vacuolar protein phytohemagglutinin targets invertase to the yeast vacuole. *Plant Cell*, **2**, 533–46.

Tanchak, M.A., Griffing, L.R., Mersey, B.G. & Fowke, L.C. (1984). Endocytosis of cationized ferritin by coated vesicles of soybean protoplasts. *Planta*, **162**, 481–6.

Tanchak, M. A., Rennie, P. J. & Fowke, L. C. (1988). Ultrastructure of the partially coated reticulum and dictyosomes during endocytosis by soybean protoplasts. *Planta*, **175**, 433–41.

Valls, L. A., Hunter, C. P., Rothman, J. H. & Stevens, T. H. (1987). Protein sorting in yeast: the localization determinant of yeast vacuolar carboxypeptidase Y resides in the propeptide. *Cell*, **48**, 887–97.

Walter, P. & Lingappa, V. (1986). Mechanism of protein translocation across the endoplasmic reticulum. *Annual Review of Cell Biology*, **2**, 499–516.

Wilkins, T. A., Bednarek, S. Y. & Raikhel, N. V. (1990). Role of propeptide glycan in post-translational processing and transport of barley lectin to vacuoles in transgenic tobacco. *Plant Cell*, **2**, 301–13.

Wilson, D. W., Wilcox, C. E., Flynn, G. C., Chen, E., Kuang, W-J., Henzel, W. J., Block, M. R., Ullrich, A. & Rothman, J. E. (1989). A fusion protein required for vesicle-mediated transport in both mammalian cells and yeast. *Nature* (*London*), **339**, 355–9.

R. L. JONES, L. STICHER AND D. S. BUSH

Secretion of hydrolases from cereal aleurone cells

The subcellular pathway and mechanisms underlying the secretion of enzymes from plant cells are not well understood (for a review, see Jones & Robinson 1989). The paucity of information about these processes stems largely from the absence of appropriate experimental systems in which to study them. As autotrophic organisms, plants do not rely on the digestion of food for their growth and development, and tissues that serve as digestive glands have evolved only in fruits, seeds and in other exceptional tissues such as the leaves of insectivorous plants. Even in the storage tissues of seeds, where mobilisation of reserves is enzymically catalysed, digestive glands are frequently not differentiated. In the case of the storage cotyledons of legumes, for example, hydrolytic enzymes are synthesised within the same cells that store the target substrates. In cotyledonary cells, substrates such as protein are stored intracellularly in protein-body vacuoles and newly synthesised digestive enzymes are targeted to this compartment rather than to the exterior of the cell (for reviews, see Chrispeels 1985; Herman 1988; and see Harris & Watson, this volume).

In this chapter we discuss the synthesis, intracellular transport and exocytosis of hydrolytic enzymes from the aleurone layer of cereal grains. This tissue is highly differentiated, functioning as a storage tissue in the mature dry grain and as a digestive gland during germination. The differentiation of the aleurone layer into a digestive gland is regulated in part by gibberellic acid (GA) and abscisic acid (ABA). During germination, GA produced by the embryo axis stimulates the aleurone layer to synthesise and secrete hydrolytic enzymes (Briggs 1973). If germination conditions are unfavourable, ABA produced by the embryo will suppress the glandular function of this tissue (Jacobsen & Chandler 1987).

The α-amylases of cereals have received considerable attention from biochemists, cell biologists and molecular biologists, and most of the experiments that we describe in this paper deal with the isozymes of this enzyme. This group of proteins is complex, consisting of many genes located on two chromosomes of the barley genome (Brown & Jacobsen 1982). Although these proteins are represented by a large complex of isozymes (see Simon & Jones 1988), there have been few attempts to adopt a standard system for naming them (see MacGregor & MacGregor 1987). In this chapter we adopt the system that identifies the principal groups of α-amylase encoded by genes on chromosome 1 and 6 according to IUPAC rules and α-amylase isozymes by their isoelectric points (pIs). The low-pI isozymes encoded on chromosome 1 are therefore referred to as BAMY1 and the high-pI isozymes encoded on chromosome 6 as BAMY2.

The secretory pathway

All secretory proteins are thought to be synthesised on the endoplasmic reticulum (ER) and transported to the cell surface via the Golgi apparatus (GApp) by secretory vesicles (Farquhar 1985). At least two pathways of intracellular transport are found in secretory animal cells, the regulated and constitutive pathways (Farquhar 1985). Secretion along the regulated pathway is a characteristic of the digestive glands of animal cells (Palade 1975). Enzymes secreted along this pathway accumulate in the cytosol and are concentrated in secretion granules. Exocytosis occurs when the cell is stimulated to secrete (Farquhar 1985). In secretion along the regulated pathway, therefore, protein synthesis and exocytosis are temporally separated.

Secretion of proteins from higher plant cells, including the secretion of hydrolytic enzymes from the aleurone layer, occurs along the constitutive pathway (for a review, see Jones & Robinson 1989). Constitutive secretion is characterised by the absence of secretory granules that concentrate secreted proteins; newly synthesised proteins are rapidly transported to the cell surface. In the barley aleurone, pulse-labelling studies show that the half-time for secretion of newly synthesised α-amylase is about 10 min (Chen & Jones 1974). Because of the rapid turnover of the pool of newly synthesised protein in the aleurone cell, the rate of secretion is closely coupled to the supply of the protein. In cells such as the cereal aleurone layer, therefore, hormones such as GA and ABA, and calcium ions, can regulate the secretion of proteins by controlling the rate of their synthesis (Fincher 1989).

Because protein synthesis and secretion are tightly coupled in constitutively secreting cells, it has proved difficult to separate these processes experimentally. Thus, any treatment or manipulation of the cell that affects protein synthesis will quickly affect the secretion of protein. This, coupled with the fact that the pool of secretory protein in the aleurone is small, has led to several proposals that secretion from cereal aleurone cells might occur by a mechanism bypassing the GApp (Jones 1972; Vigil & Ruddat 1973). More recent experiments with the sodium ionophore monensin (Mon) have established the involvement of the GApp in the intracellular transport of α-amylase in barley aleurone (Heupke & Robinson 1985; Melroy & Jones 1986). Treatment of barley aleurone layers with Mon inhibits only the secretion of α-amylase. α-Amylase that is synthesised in Mon-treated aleurone layers accumulates in the cytosol in a GApp-derived vesicle (Melroy & Jones 1986).

In barley aleurone and rice scutellum Mon does not completely inhibit the secretion of α-amylase (Mitsui *et al.* 1985; Melroy & Jones 1986). In both rice and barley two groups of α-amylase isozymes are synthesised and secreted (for a review, see Akazawa & Hara-Nishimura 1985), and Mon affects the secretion of these two groups of enzymes differently. In rice scutellum tissue Mon inhibited the secretion of only the R-form of α-amylase, the S-form of the enzyme being much less affected by the ionophore (Mitsui *et al.* 1985). In barley Mon affects the secretion of HAMY2 α-amylase isozymes more than the HAMY1 forms (Melroy & Jones 1986). To account for the differential effects of Mon on α-amylase secretion in rice scutellum Akazawa & Hara-Nishimura (1985) proposed two hypotheses, one of which was a route of secretion for R-forms of the enzyme along a Mon-insensitive pathway that bypassed the GApp.

Immunocytochemical evidence does not favour a two-route model for the secretion of the two groups of barley α-amylase isozymes. Using antibodies raised against HAMY1 and HAMY2, Zingen-Sell *et al.* (1990) showed that the secretion of both isozyme groups from barley aleurone cells occurred via the GApp. These experiments confirmed the immunocytochemical observations of Gubler *et al.* (1986), who used antibodies to a mixture of HAMY1 and HAMY2 to show the involvement of the ER and GApp in the synthesis and transport of barley α-amylase.

Signal sequences and targeting to the ER

The amino terminus of proteins secreted from barley aleurone possesses a signal peptide that is indistinguishable in both structure and function from signal sequences isolated from other eukaryotic or prokaryotic organisms (see von Heijne 1984). After studying the amino terminal region of 134 eukaryotic and 39 prokaryotic sequences, von Heijne (1984) formulated rules that described all signal sequences. Signal peptides possess a positively charged amino terminus, a core of hydrophobic amino acid residues and a carboxyl terminus of about five amino acid residues that are relatively polar (von Heijne 1985). A comparison of the N-terminal

```
+     + +
M A N K H L S|L S L F L V L L|G L S A S L A S G

+         + +
M D Y Y R K Y|A A I F L V T L S V F L|H V L H S
```

Fig. 1. A comparison of the amino acid sequences of a HAMY2 isozyme derived from cDNA clones pHV19 (Chandler *et al.* 1984) and pM/C (Rogers 1985) (upper sequence) and human chorionic gonadotropin, α-subunit (von Heijne 1984) (lower sequence). The amino acid residues that constitute the hydrophobic core of the signal sequence are enclosed by a box. Positively charged amino acid residues at the N terminus are indicated (+).

sequence of a HAMY2 isozyme of barley α-amylase encoded by cDNA clones pHV19 (Chandler *et al.* 1984) or pM/C (Rogers 1985) with the signal sequence of a typical secretory protein (α-subunit of human chorionic gonadotropin; von Heijne 1984) illustrates the marked structural similarities among presumed signal sequences (Fig. 1).

That the N-terminal segment of the cereal α-amylase molecule serves as a signal sequence is indicated by several types of experiment. Okita *et al.* (1979), Mozer (1980) and Boston *et al.* (1982) have shown that α-amylase synthesised *in vitro* is about 2 kD larger than the secreted form of the enzyme (see also Fig. 2). Addition of microsomal membranes isolated from dog pancreas (K. Aoyagi & R.L. Jones, unpublished results) or barley aleurone (L. Sticher & R.L. Jones, unpublished results) will bring about the cotranslational modification of barley α-amylase *in vitro*. The protein that is processed *in vitro* is indistinguishable from the protein that is secreted from aleurone cells. The microsomal membranes isolated from barley aleurone will also bring about the cotranslational processing of other plant proteins that are synthesised *in vitro*. Thus, we have shown that tomato protease inhibitor I synthesised by the wheat germ *in vitro* system can be cotranslationally modified by the addition of barley aleurone microsomal membranes (K.W. Osteryoung, L. Sticher, A. Bennett & R.L. Jones, unpublished results). However, barley microsomal membranes are more efficient than dog pancreas microsomes in processing tomato protease inhibitor I *in vitro*.

The synthesis and secretion of α-amylase in several different heterologous systems also establishes that the signal sequence of cereal α-amylases is important in targeting this enzyme to the ER (Gatenby *et al.* 1986; Rothstein *et al.* 1987; Simon & Jones, 1988; Aoyagi *et al.* 1990). We have studied the synthesis and secretion of barley α-amylase from

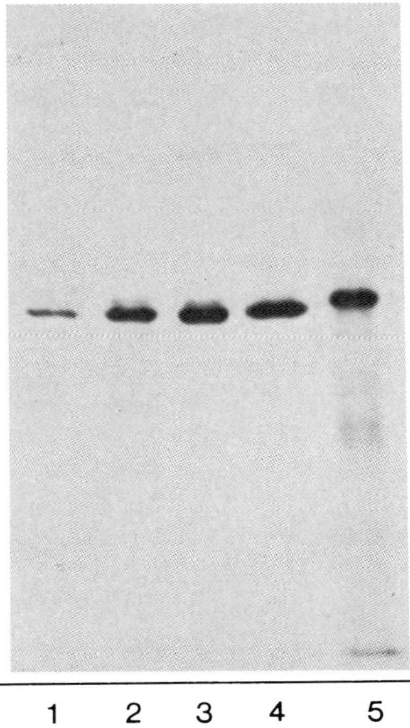

Fig. 2. Sodium dodecyl sulphate/polyacrylamide gel electrophoresis and fluorography of radiolabelled proteins immunoprecipitated with anti-α-amylase antibody from incubation media (lanes 1, 3) and extracts (lanes 2, 4) of *Xenopus* oocytes (lanes 1, 2) programmed with barley aleurone poly(A)$^+$-RNA and pulse-labelled with [^{35}S]methionine, and barley aleurone layers (lanes 3, 4) pulse-labelled with [^{35}S]methionine. Lane 5: immunoprecipitated radiolabel in α-amylase synthesised *in vitro* by the wheat germ *in vitro* synthesising system. (From Simon & Jones 1988, with permission.)

Xenopus laevis oocytes injected with either poly(A)$^+$ RNA (Simon & Jones 1988) or with recombinant plasmids containing cDNA encoding HAMY1 and HAMY2 isozymes (Simon & Jones 1988; Aoyagi *et al.* 1990). *Xenopus* oocytes programmed with poly(A)$^+$ RNA secrete catalytically active α-amylase along a Mon-sensitive and tunicamycin-insensitive pathway that involves the GApp (Fig. 3; Simon & Jones 1988). The molecular weights, isoelectric points and antigenic properties of the α-amylases secreted by oocytes programmed with RNA or α-amylase cDNA are indistinguishable from those of native secreted α-amylase, showing that the post-translational processing of the amylase

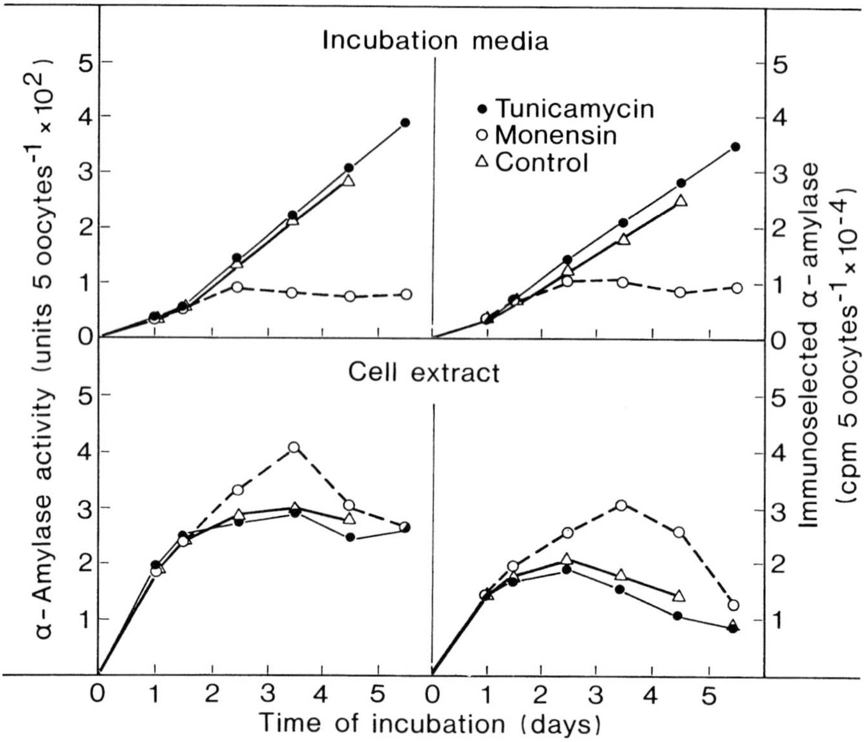

Fig. 3. Time course of α-amylase secretion and accumulation in *Xenopus* oocytes programmed with barley aleurone poly(A)$^+$-RNA. Oocytes were pulse-labelled by incubation in [^{3}H]leucine from time 0, and either tunicamycin was added at time 0 or monensin was added 24 h later. α-Amylase activity (left panels) and immunoprecipitable radioactivity (right panels) were determined by analysing portions of the medium and cell extract. (From Simon & Jones 1988, with permission.)

molecule is similar in oocyte and aleurone cytoplasm (Figs. 2 and 4; Simon & Jones 1988; Aoyagi *et al.* 1990).

Post-translational protein modification

Calcium binding

Barley α-amylase is a calcium-containing metalloprotein, and all isozymes bind at least one Ca^{2+} per molecule (Bush *et al.* 1989*b*). Calcium binding is essential for the catalytic activity of amylase, and it confers stability

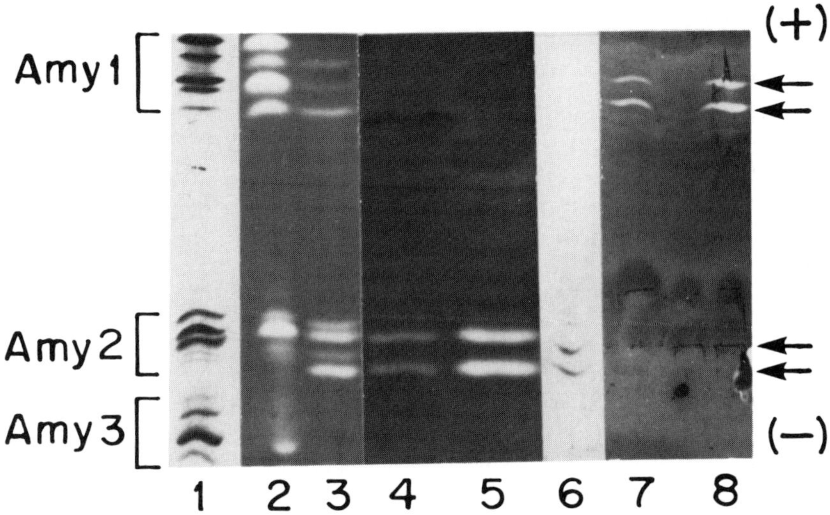

Fig. 4. Isoelectric focusing of proteins from homogenates and incubation media of *Xenopus* oocytes injected with barley RNA and DNA. α-Amylase in incubation media of oocytes injected with poly(A)$^+$RNA (lane 3), pSP.M/C (lane 4), pSP.1–28 (lanes 5 and 6), and pSP.E (lane 7), and homogenate of oocytes injected with pSP.E (lane 8). Native α-amylase containing HAMY1 (Amy1), HAMY2 (Amy2) and a HAMY2/amylase inhibitor complex (Amy3) were used as a standard (lanes 1 and 2). Protein was detected by fluorography (lane 1), enzyme activity (lanes 2–5) and protein immunoblotting (lane 6). Incubation media of homogenates of oocytes injected with RNA (lane 3) or DNA (lanes 4–8) were collected after 2–4 days of incubation and 25 μl were loaded per lane. The positions of the amylase isozymes synthesised by the oocytes injected with plasmids are indicated by arrows. (From Aoyagi *et al.* 1990, with permission.)

against proteolytic degradation (Bush *et al.* 1989*b*). Because the ER of barley aleurone cells contains precursor forms of the α-amylase molecule that are biologically active, we presume that the binding of Ca^{2+} to the amylase molecule is an early event in the process of enzyme synthesis.

We have studied the structure of the barley α-amylase molecule by analysing the nucleotide sequences of HAMY1 and HAMY2 forms using the approach described by Chou & Fassman (1978). We have also compared the predicted secondary structure of HAMY obtained using Chou–Fassman rules with the structure of porcine pancreatic amylase (PPA) determined by X-ray crystallography (Buisson *et al.* 1987). This comparison confirms that similarities between HAMY and PPA at the amino acid level (Rogers 1985) are also found at the level of secondary structure (Bush *et al.* 1989*b*). In particular three of the four predicted Ca^{2+}-binding amino acid residues of PPA (Asn100, Asp159, Asp167, His201) are in structurally similar positions in HAMY (Asn92, Asp138, Asp149). There is no structural analogue to the His201 residue of PPA in HAMY (Bush *et al.* 1989*b*). We propose that because the Ca^{2+}-binding domain of HAMY is located near the N-terminus of the molecule, Ca^{2+} is bound to the enzyme as it enters the lumen of the ER, conferring catalytic activity on precursor forms of the enzyme.

Calcium binding to HAMY *in vitro* is an irreversible process. When affinity purified HAMY is incubated at pCa 7, catalytic activity is lost and cannot be recovered by the addition of Ca^{2+} (Bush *et al.* 1989*b*). This contrasts with the reversible inactivation of animal and bacterial α-amylases (Vallee *et al.* 1959). On the basis of our structural comparisons of HAMY and PPA we propose that the differences in Ca^{2+} reactivation between these enzymes are related to differences in the location and function of Cys residues. In PPA, Cys70 and Cys115 form a disulphide bridge that binds domains A and B, forming the Ca^{2+} binding site. When Ca^{2+} is removed from PPA, the Ca^{2+} binding amino acid residues continue to be held in place by the disulphide bond between residues 70 and 115. These Cys residues are absent from HAMY, however, and domains A and B are held together by Ca^{2+}. When Ca^{2+} is removed from HAMY, domains A and B are free to move away from each other. We therefore speculate that the binding of Ca^{2+} to the HAMY molecule is not a spontaneous process, but rather that it is catalysed by a protein resident in the ER, such as BiP (binding protein; for a review, see Pelham 1989).

Calcium binding is also important for the stability of the HAMY molecule. We have shown that removal of Ca^{2+} by incubation of the purified enzyme at pCa 7 makes the enzyme susceptible to proteolytic degradation by chymotrypsin. The susceptibility of α-amylase to proteolytic attack in the absence of bound Ca^{2+} also strongly supports the idea that Ca^{2+} binding to the enzyme is an early event in its synthesis.

We have demonstrated that the activity of α-amylase in the lumen of the ER of barley aleurone is dependent upon the maintenance of high levels of Ca^{2+} in this compartment. Isolated ER vesicles containing active α-amylase were incubated in pCa 4 or pCa 7 in the presence or absence of the Ca ionophores A23187 or ionomycin (Bush *et al.* 1989*b*). When vesicles were incubated in pCa 7 in the presence of ionophores, the activity of α-amylase in the lumen was markedly reduced relative to that of vesicles incubated in pCa 7 in the absence of ionophore.

We have shown that the lumen of the ER of barley aleurone contains higher levels of Ca^{2+} than does the cytosol. Using the Ca^{2+}-sensitive dye indo-1, Bush & Jones (1987, 1988) found cytosolic Ca^{2+} to be about 0.3 μM in protoplasts incubated in GA + 20 mM Ca^{2+}. Calcium levels were homeostatically maintained at about 0.3 μM in protoplasts for up to 36 h of incubation in GA and 20 mM Ca^{2+} (Bush & Jones 1988). Endoplasmic reticulum isolated from protoplasts incubated in GApp plus 20 mM Ca^{2+} and loaded with indo-1 also contained indo-1, thus permitting the measurement of Ca^{2+} in this compartment (Bush *et al.* 1989*a*). Fluorescence from indo-1 in isolated ER fractions showed that Ca^{2+} was present in the ER lumen at a concentration of at least 3 μM. This value for lumenal ER Ca^{2+} levels represents a low estimate since indo-1 is useful for measuring Ca^{2+} only in the range 0.05–3.0 μM (Grynkiewicz *et al.* 1985). More recently we have estimated that isolated ER vesicles can accumulate as much as 1 mM Ca^{2+} by the activity of a Ca^{2+} located in the ER membrane (D.S. Bush & R.L. Jones, unpublished results).

Elevated ER Ca^{2+} levels are maintained by a Ca^{2+} transporter located in the ER membrane (Bush *et al.* 1989*a*). This transporter is an ATPase whose activity is inhibited by vanadate but not by nitrate. The K_m of the ATPase for Ca^{2+} is 0.5 μM, a value consistent with its role in the transport of Ca^{2+} from the aleurone cytosol to the ER. The activity of the ER Ca^{2+} ATPase is regulated by GA (Fig. 5) and ABA. The activity of the ER ATPase increases with the duration of incubation of aleurone layers or protoplasts in GA. Thus, the activity of the transporter is barely detectable in ER isolated from freshly isolated aleurone, but after about 4–6 h in GA it rises relative to control layers incubated in water, reaching a maximum after about 24 h (Bush *et al.* 1989*a*). Incubation in ABA, on the other hand, lowers Ca^{2+} transport activity in the ER to levels lower than that found in water controls, and addition of ABA to aleurone layers preincubated in GA reverses the effects of GA (Bush & Jones 1990). The effects of GA and ABA on Ca^{2+} transport in the ER are not mediated via changes in the amount of ER membrane. Rather, the effects of these two growth regulators are to alter the concentration of transporter in the membrane (Bush *et al.* 1989*a*; D.S. Bush & R.L. Jones, unpublished results).

The kinetics of the GA-induced increase in ER Ca^{2+} transport activity are almost identical with the kinetics of GA-induced α-amylase synthesis (Fig. 5). From these data and others showing the dependence of α-amylase synthesis in barley aleurone on added Ca^{2+} (Jacobsen *et al.* 1970; Jones & Jacobsen 1983), we conclude that GA-induced Ca^{2+} transport into the ER is an important step in the regulation of α-amylase synthesis in the aleurone cell.

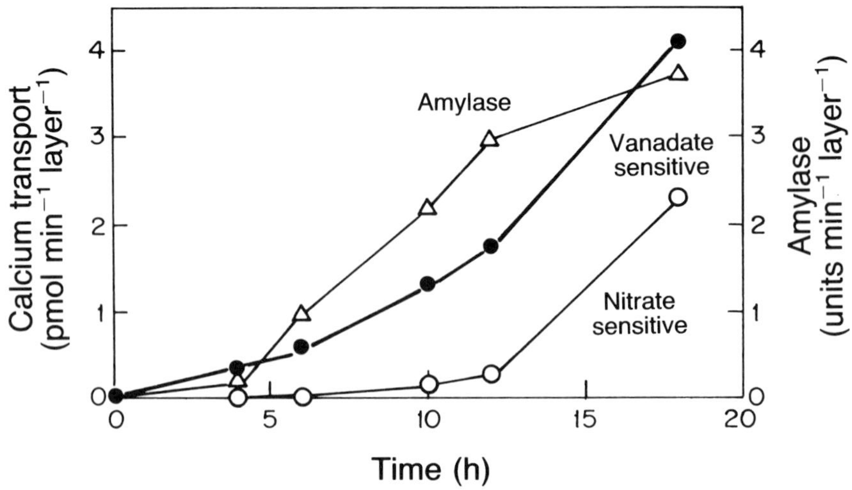

Fig. 5. Comparison of the time courses of GA-induced α-amylase secretion and Ca^{2+}-transport into the ER (vanadate sensitive) and the vacuole (nitrate sensitive). Calcium transport and α-amylase activity were measured on membranes obtained from aleurone layers that had been incubated for 4, 6, 10, 12, and 18 h in a medium containing GA and Ca^{2+}. (From Bush & Jones 1990, with permission.)

Isoelectric points

Following Ca^{2+} binding, α-amylase isozymes are further post-translationally modified by a reaction(s) that causes a change in the pI of isozymes but does not involve a significant change in enzyme mass (Jacobsen *et al.* 1988). This post-translational modification was discovered by Jacobsen *et al.* (1985) while studying the synthesis and secretion of α-amylase from protoplasts isolated from aleurone layers of the Himalaya cultivar of barley. They found more enzymically active isozymes of HAMY1 in homogenates of protoplasts than were secreted into the incubation medium, and they proposed that the additional isozymes found within the cell were precursor forms of enzymes that were secreted (Jacobsen *et al.* 1985).

Evidence that precursor forms of α-amylase gave rise to new isozymes having lower pI values came from pulse-chase labelling experiments (Jacobsen *et al.* 1988). Radiolabel from [^{35}S]methionine was first incorporated into two HAMY1 isozymes with pI values of 4.9 and 4.65, and, during a subsequent chase with unlabelled methionine, radioactivity appeared in isozymes having pI values of 4.72 and 4.56. By comparing the pI values of the HAMY1 isozymes of Himalaya barley with those of the Betzes cultivar of barley, which has only HAMY1 pI values 4.9 and 4.72, Jacobsen *et al.* (1988) surmised that HAMY1 pI 4.9 was a precursor of pI 4.72 and that pI 4.65 was a precursor of pI 4.56.

The precise relationship between the various precursor and product α-amylase isozymes of barley was established by expressing cDNAs encoding barley amylases in *Xenopus* oocytes. Three cDNA clones, one encoding HAMY1 (clone E, Rogers & Milliman 1984) and two encoding HAMY2 (clones pM/C, Rogers 1985, and clone 1–28, Deikman & Jones 1985) were subcloned into the plasmid sP6 (Aoyagi *et al.* 1990). The recombinant plasmids pSP.E, pSP.M/C and pSP.1–28 were injected into oocyte nuclei and transcription of the α-amylase DNAs was monitored by the hybridisation of ^{32}P-labelled 1–28 cDNA to RNA isolated from the oocytes (Aoyagi *et al.* 1990). Oocytes injected with recombinant plasmids also secreted catalytically active α-amylases into the incubation medium (Fig. 4). Two enzymically active α-amylases isozymes were secreted from oocytes injected with each of the recombinant plasmids (Fig. 4). These isozymes were indistinguishable from native HAMY1 or HAMY2 in their isoelectric points. Thus, when pSP.E was injected into the oocyte, two isozymes with pI values 4.9 and 4.72 accumulated in the incubation medium. These isozymes correspond to the precursor (pI 4.9) and product (pI 4.7) forms of HAMY1. Two isozymes with pI values of 6.1 and 6.0 were secreted from the oocyte when pSP.M/C and pSP.1–28 were injected (Fig. 2). We presume that isozyme pI 6.1 is the precursor of pI 6.0. This presumption is based on several observations. First, because two proteins are synthesised from a single cDNA, one must have arisen from post-translational modification. Second, isozyme pI 6.0 accumulates in barley aleurone incubation medium to high levels, and isozyme pI 6.1 is found only within the aleurone cell and at low levels. Third, by analogy with HAMY1 isozymes, if the post-translational modification involves a lowering of pI, then isozyme 6.1 is the precursor of pI 6.0.

Because precursor and product HAMY can be separated by isoelectric focusing and identified by their enzymic activities, we studied the subcellular location of the post-translational modification step. Using density gradient centrifugation methods we have separated a light membrane fraction enriched in the GApp marker enzyme inosine diphophatase (IDPase) from a fraction that possesses IDPase as well as the ER marker cytochrome *c* reductase. Whereas the ER/GApp fraction contains a higher proportion of precursor forms of HAMY, the light GApp fraction contains more of the secreted form of the enzyme (L. Sticher & R.L. Jones, unpublished results). From these data we conclude that post-translational modification of HAMY begins in the ER and processing continues as HAMY is transported through the endomembrane system. The light membrane fraction that possesses IDPase activity is probably the secretory vesicle because it is enriched in isozyme forms that accumulate in the incubation medium.

Neither the nature of the post-translational modification that lowers α-amylase pI nor its function are known. We have eliminated several possible modifications that might cause a change in enzyme charge without significantly altering molecular weight. Thus, fatty acylation, phosphorylation and sulphation do not occur, and these proteins are not *N*- or *O*-glycosylated (Jacobsen *et al.* 1988). We have recently isolated an enzyme from barley aleurone cells that will lower the pI of precursor forms of HAMY1 *in vitro* (L. Sticher & R. L. Jones, unpublished results). This enzyme will also lower the pI of human salivary α-amylase. Because human salivary amylases are post-translationally modified by the deamidation of glutamine (Karn *et al.* 1974), we propose that a deamidation step is also involved in the lowering of HAMY pI. The deamidation of Gln is known to occur post-translationally in other proteins and this reaction may occur frequently (for a review, see Wold 1985).

Few clues are available on the function of this post-translational modification of HAMY. Experimental evidence indicates that it is not related to the transport of α-amylase to the cell exterior. Thus, in barley aleurone, precursor forms of HAMY1 are found in the incubation medium, and, in the case of *Xenopus* oocytes injected with RNA or DNA encoding α-amylase, both precursor and product α-amylase are secreted in approximately equal amounts. We favour the view that secretion of HAMY occurs along the bulk flow or default pathway (see Chrispeels & Tague 1990).

Other post-translational modifications

Although the α-amylases secreted by the barley aleurone are not glycoproteins, many of the other hydrolases secreted by this tissue are post-translationally modified by the addition of sugar moieties (Fincher 1989). A mixed linked β-glucanase (Fincher *et al.* 1986) and a nuclease I (Brown & Ho 1986, 1987) are glycosylated. In the case of the nuclease I examined in detail by Brown & Ho (1987) an Asn-linked high mannose oligosaccharide is linked to the secreted protein, but the subcellular site of this glycosylation step has not been examined (for a recent review of post-translational protein modification in plants, see Chrispeels & Tague 1990).

An endoprotease (EP-A) secreted by barley aleurone layers (Koehler & Ho 1988) is synthesised as a proenzyme and the mature enzyme arises by proteolytic processing (S. Koehler & D. Ho, unpublished results). The steps in the processing of EP-A may be similar to the processing of aleurain, a cysteine protease synthesised by barley aleurone cells but targeted to a subcellular lysosome-like compartment (B. Holwerda, N.J. Galvin & J.C. Rogers, unpublished results). In the case of aleurain, a 42 kD, glycosylated proenzyme is processed to the 32 kD mature form by removal of a 10 kD pro sequence. Although the subcellular site of proteolytic processing of these protease proenzymes is not known, there is evidence that in cereals the pro sequence is cleaved from the secreted proforms of proteases extracellularly. Such appears to be the case for the carboxypeptidase I secreted from barley scutellum (Doan & Fincher 1988) and for EP-A secreted from the aleurone layer (S. Koehler & D. Ho, unpublished results). As is the case for the post-translational modification of secreted α-amylase, the role of the processing of other secreted proteins is not known.

Prospects

Although progress in understanding how proteins are synthesised and transported from cereal aleurone has been slow, several recent developments are likely to increase our knowledge of these processes. In particular, the ability to transform cereal aleurone protoplasts will provide the means of examining the role of particular amino acid sequences in the targeting of proteins to various compartments in the aleurone cell (Huttly & Baulcombe 1989). The expression of genes for aleurone secretory proteins in heterologous systems will provide another means for examining the role of specific sequences in other cytoplasmic backgrounds (Gatenby *et al.* 1986; Rothstein *et al.* 1987; Aoyagi *et al.* 1990). The ability to express wheat aleurone α-amylase genes in *Saccharomyces cerevisiae* (Rothstein *et al.* 1987) raises the possibility of using the wide range of secretory mutants that exist in yeast (Schekman 1985) to study the steps in the intracellular transport of aleurone proteins.

Research on the intracellular transport and secretion of enzymes in cereals may also lead to new concepts and discoveries. For example, the observation that the thiol protease aleurain is similar to the rat liver lysosomal protease cathepsin H (Rogers *et al.* 1985) and is targeted to a subcellular compartment in barley aleurone that is structurally similar to animal lysosomes (B. Holwerda, N.J. Galvin & J.C. Rogers, unpublished results) may lead to the discovery of a *bona fide* lysosome in plants.

Summary

Enzyme secretion from cereal aleurone occurs along the constitutive pathway. This route of secretion is characterised by close coupling between the synthesis of a secreted protein and its rate of exocytosis. In cells such as those of cereal aleurone, enzyme secretion can be controlled at the level of enzyme synthesis by the plant growth regulators GA and ABA, which control the expression of the genes for secretory proteins. The secreted proteins of cereal aleurone possess signal sequences that direct them to the membrane of the ER. These signal sequences are similar in structure and function to the signal peptides found in other organisms. The synthesis of α-amylase having a signal peptide and cleavage of this peptide by microsomal membranes from barley has now been demonstrated *in vitro*. After removal of the signal peptide, barley α-amylases are activated and stabilised by the binding of Ca^{2+}. Calcium binding is an early event in the synthesis of the mature active enzyme, and it occurs in the lumen of the ER. The Ca^{2+} that binds to α-amylase is pumped into the lumen of the ER by an ATPase whose activity is also regulated by GA and ABA. Active α-amylase in barley undergoes another post-translational modification prior to secretion. This modification occurs in the endomembrane system and causes a lowering of the pI of the enzyme. The function of this post-translational modification is not known, but circumstantial evidence shows that it is not involved in the targeting of the enzyme to the cell surface. Our evidence supports the proposal that aleurone α-amylase and other secreted hydrolases do not contain unique targeting signals and are therefore transported to the cell surface along the default pathway.

References

Akazawa, T. & Hara-Nishimura, I. (1985). Topographic aspects of biosynthesis, extracellular secretion, and intracellular storage of proteins in plant cells. *Annual Review of Plant Physiology and Plant Molecular Biology*, **36**, 441–72.

Aoyagi, K., Sticher, L., Wu, M. & Jones, R.L. (1990). The expression of barley α-amylase genes in *Xenopus laevis* oocytes. *Planta*, **180**, 333–40.

Boston, R.S., Miller, T.J., Mertz, J.E. & Burgess, R.R. (1982). In vitro synthesis and processing of wheat α-amylase. *Plant Physiology*, **69**, 150–4.

Briggs, D.E. (1973). Hormones and carbohydrate metabolism in germinating cereal grains. In *Biosynthesis and its Control in Plants*.

Annual Proceedings of the Phytochemical Society, No. 9, pp. 219–77. New York: Academic Press.

Brown, P.H. & Ho, T.-H.D. (1986). Barley aleurone layers secrete a nuclease in response to gibberellic acid. Purification and partial characterization of the associated ribonuclease, deoxyribonuclease, and 3′ nucleotidase activities. *Plant Physiology*, **82**, 801–6.

Brown, P.H. & Ho, T.-H.D. (1987). Biochemical properties and hormonal regulation of barley nuclease. *European Journal of Biochemistry*, **168**, 357–64.

Brown, A.H.D. & Jacobsen, J.V. (1982). Genetic basis and natural variation of α-amylase isozymes in barley. *Genetics Research*, **40**, 315–24.

Buisson, G., Duée, E., Haser, R. & Payan, F. (1987). Three dimensional structure of porcine pancreatic α-amylase at 2.9 A resolution. Role of calcium in structure and activity. *EMBO Journal*, **6**, 3909–16.

Bush, D.S., Biswas, A.K. & Jones, R.L. (1989*a*). Gibberellic-acid-stimulated Ca^{2+} accumulation in endoplasmic reticulum of barley aleurone: Ca^{2+} transport and steady state levels. *Planta*, **178**, 411–20.

Bush, D.S. & Jones, R.L. (1987). Measurement of cytoplasmic calcium in aleurone protoplasts using indo-1 and fura-2. *Cell Calcium*, **8**, 455–72.

Bush, D.S. & Jones, R.L. (1988). Cytoplasmic calcium and α-amylase secretion from barley aleurone protoplasts. *European Journal of Cell Biology*, **46**, 466–9.

Bush, D.S. & Jones, R.L. (1990). Hormonal regulation of Ca^{2+} transport in microsomal vesicles isolated from barley aleurone layers. In *Calcium in Plant Growth and Development*, ed. R.T. Leonard & P.K. Hepler, pp. 60–6. Bethesda, MD: American Society of Plant Physiologists, Bethesda.

Bush, D.S., Sticher, L., Van Huystee, R., Wagner, D. & Jones, R.L. (1989*b*). The calcium requirement for stability and enzymatic activity of two isoforms of barley aleurone α-amylase. *Journal of Biological Chemistry*, **264**, 19 392–8.

Chandler, P.M., Zwar, J.A., Jacobsen, J.V., Higgins, T.J.V. & Inglis, S. (1984). The effect of gibberellic acid and abscisic acid on α-amylase mRNA levels in barley aleurone layers. Studies using an α-amylase cDNA clone. *Plant Molecular Biology*, **3**, 407–18.

Chen, R. & Jones, R.L. (1974). Studies on the release of barley aleurone cell proteins: kinetics of labelling. *Planta*, **119**, 193–206.

Chou, A.Y. & Fassman, G.D. (1978). Empirical predictions of protein conformation. *Annual Review of Biochemistry*, **47**, 251–76.

Chrispeels, M.J. (1985). The role of the Golgi apparatus in the transport and the post-translational modifications of vacuolar (protein body) proteins. *Oxford Surveys in Plant Molecular Cell Biology*, **2**, 43–68.

Chrispeels, M.J. & Tague, B.W. (1990). Protein sorting in the secretory system of plant cells. *International Review of Cytology*, (in press).

Deikman, J. & Jones, R. L. (1985). Control of α-amylase mRNA accumulation by gibberellic acid and calcium in barley aleurone layers. *Plant Physiology*, **78**, 192–8.

Doan, N.P. & Fincher, G.B. (1988). The A- and B-chains of carboxypeptidase I from germinated barley originate from a single precursor polypeptide. *Journal of Biological Chemistry*, **263**, 11 106–10.

Farquhar, M.G. (1985). Progress in unravelling pathways of Golgi traffic. *Annual Review of Cell Biology*, **1**, 447–88.

Fincher, G.B. (1989). Molecular and cellular biology associated with endosperm mobilization in germinating cereal grains. *Annual Review of Plant Physiology and Plant Molecular Biology*, **40**, 305–46.

Fincher, G.B., Lock, P.A., Morgan, M.M., Lingelbach, K., Wettenhall, R.E.H., Mercer, J.F.B., Brandt, A. & Thomsen, K.K. (1986). Primary structure of the (1→3, 1→4)-β-D-glucan 4-glucohydrolase from barley aleurone. *Proceedings of the National Academy of Sciences, USA*, **83**, 2081–5.

Gatenby, A.A., Boccara, M., Baulcombe, D.C. & Rothstein, S.J. (1986). Expression of a wheat α-amylase gene in *Escherichia coli*: recognition of the translational initiation site and the signal peptide. *Gene*, **45**, 11–18.

Grynkiewicz, G., Poenie, M. & Tsien, R.Y. (1985). A new generation of Ca^{2+} indicators with greatly improved fluorescence properties. *Journal of Biological Chemistry*, **260**, 3440–50.

Gubler, F., Jacobsen, J.V. & Ashford, A.E. (1986). Involvement of the Golgi apparatus in the secretion of α-amylase from gibberellin-treated barley aleurone cells. *Planta*, **168**, 447–52.

Herman, E.M. (1988). Immunocytochemical localization of macromolecules with the electron microscope. *Annual Review of Plant Physiology and Plant Molecular Biology*, **39**, 139–55.

Heupke, H.-J. & Robinson, D.G. (1985). Intracellular transport of α-amylase in barley aleurone cells: evidence for the participation of the Golgi apparatus. *European Journal of Cell Biology*, **39**, 265–72.

Huttly, A.K. & Baulcombe, D.C. (1989). A wheat α-Amy2 promoter is regulated by gibberellin in transformed oat aleurone protoplasts. *EMBO Journal*, **8**, 1907–13.

Jacobsen, J.V., Bush, D.S., Sticher, L. & Jones, R.L. (1988). Evidence for precursor forms of the low isoelectric point α-amylase isozymes secreted by barley aleurone cells. *Plant Physiology*, **88**, 1168–74.

Jacobsen, J.V. & Chandler, P.M. (1987). Gibberellin and abscisic acid in germinating cereals. In *Plant Hormones and Their Role in Plant Growth and Development*, ed. P.J. Davies, pp. 164–93. Dordrecht: Martinus Nijhoff.

Jacobsen, J.V., Scandalios, J.G. & Varner, J.E. (1970). Multiple forms of amylase induced by gibberellic acid in isolated barley aleurone layers. *Plant Physiology*, **45**, 367–71.

Jacobsen, J.V., Zwar, J.A. & Chandler, P.M. (1985). Gibberellic-acid-responsive protoplasts from mature aleurone of Himalaya barley. *Planta*, **163**, 430–8.

Jones, R. L. (1972). Fractionation of the enzymes of the barley aleurone layer: Evidence for a soluble mode of enzyme release. *Planta*, **103**, 95–109.

Jones, R. L. & Jacobsen, J. V. (1983). Calcium regulation of the secretion of α-amylase isoenzymes and other proteins from barley aleurone layers. *Planta*, **158**, 1–9.

Jones, R. L. & Robinson, D. G. (1989). Protein secretion in plants. *New Phytologist*, **111**, 567–97.

Karn, R. C., Rosenblum, B. B., Ward, J. C., Merritt, A. D. & Shulkin, J. D. (1974). Immunological relationships and post-translational modifications of human salivary amylase (Amy_1) and pancreative amylase (Amy_2) isozymes. *Biochemical Genetics*, **12**, 485–99.

Koehler, S. & Ho, T.-H. D. (1988). Purification and characterization of gibberellic acid-induced cystein endoproteases in barley aleurone layers. *Plant Physiology*, **87**, 95–103.

MacGregor, E. A. & MacGregor, A. W. (1987). Studies of cereal α-amylase using cloned DNA. *CRC Critical Reviews in Biotechnology*, **5**, 129–42.

Melroy, D. & Jones, R. L. (1986). The effect of monensin on intracellular transport and secretion of α-amylase isoenzymes in the barley aleurone. *Planta*, **167**, 252–9.

Mitsui, T., Akazawa, T., Christeller, J. T. & Tartakoff, A. (1985). Biosynthesis of rice seed α-amylase. Two pathways of amylase secretion by the scutellum. *Archives of Biochemistry and Biophysics*, **241**, 315–28.

Mozer, T. J. (1980). Partial purification and characterization of the mRNA for α-amylase from barley aleurone layers. *Plant Physiology*, **65**, 834–7.

Okita, T. W., DeCaleys, R. & Rappaport, L. (1979). Synthesis of a possible precursor of α-amylase in wheat aleurone cells. *Plant Physiology*, **63**, 195–200.

Palade, G. (1975). Intracellular aspects of the process of protein secretion. *Science*, **189**, 347–58.

Pelham, H. R. B. (1989). Control of protein exit from the endoplasmic reticulum. *Annual Review of Cell Biology*, **5**, 1–23.

Rogers, J. C. (1985). Two barley α-amylase gene families are regulated differently in aleurone cells. *Journal of Biological Chemistry*, **260**, 3731–8.

Rogers, J. C., Dean, D. & Heck, G. R. (1985). Aleurain: a barley thiol protease closely related to mammalian cathepsin H. *Proceedings of the National Academy of Sciences, USA*, **82**, 6512–16.

Rogers, J. C. & Milliman, C. (1984). Coordinate increase in major transcripts from the high pI α-amylase multigene family in barley aleurone cells stimulated with gibberellic acid. *Journal of Biological Chemistry*, **259**, 12 234–40.

Rothstein, S.J., Lahners, K.N., Lazarus, C.M., Baulcombe, D.C. & Gatenby, A.A. (1987). Synthesis and secretion of wheat α-amylase in *Saccharomyces cerevisiae*. *Gene*, **55**, 353–6.

Schekman, R. (1985). Protein localization and membrane traffic in yeast. *Annual Review of Cell Biology*, **1**, 115–43.

Simon, P. & Jones, R.L. (1988). Synthesis and secretion of catalytically active barley α-amylase isoforms by *Xenopus* oocytes injected with barley mRNAs. *European Journal of Cell Biology*, **47**, 213–21.

Vallee, B.L., Stein, E.A., Sumerwell, W.N. & Fischer, E.H. (1959). Metal content of α-amylase of various origins, *Journal of Biological Chemistry*, **234**, 2901–5.

Vigil, E.L. & Ruddat, M. (1973). Effect of gibberellic acid and actinomycin D on the formation and distribution of rough endoplasmic reticulum in barley aleurone cells. *Plant Physiology*, **51**, 549–58.

von Heijne, G. (1984). Analysis of the distribution of charged residues in the N-terminal region of signal sequences: implications for protein export in prokaryotic and eukaryotic cells. *EMBO Journal*, **3**, 2315–18.

von Heijne, G. (1985). Signal sequences. The limits of variation. *Journal of Molecular Biology*, **184**, 99–105.

Wold, F. (1985). Reactions of amide side-chains of glutamine and asparagine in vivo. *Trends in Biochemical Sciences*, **10**, 4–6.

Zingen-Sell, I., Hillmer, S., Robinson, D.G. & Jones, R.L. (1990). Localization of α-amylase isozymes within the endomembrane system of barley aleurone. *Protoplasma*, **154**, 16–24.

L. A. STAEHELIN, T. H. GIDDINGS, S. LEVY, M. A. LYNCH, P. J. MOORE AND K. M. M. SWORDS

Organisation of the secretory pathway of cell wall glycoproteins and complex polysaccharides in plant cells

The Golgi apparatus of plant cells differs from its animal counterpart in two major ways: its structural organisation and the nature of its secretory products. The term Golgi apparatus (GApp) refers to the complement of Golgi stacks (dictyosomes) within a eukaryotic cell. These stacks are interconnected by tubular elements and are clustered adjacent to the nucleus in animal cells, but dispersed throughout the cytoplasm in plant cells. Plant Golgi stacks consist usually of five to eight flattened cisternae and display a distinct morphological polarity (Robinson & Kristen 1982).

Extensive biochemical and immunocytochemical evidence indicates that the structural polarity of Golgi stacks reflects an underlying functional polarity. This has been most clearly shown for mammalian cells, where a series of specific reactions involved with the processing of *N*-linked glycoproteins have been localised to successive cisternae such that processing can progress in an ordered manner. Thus, newly synthesised proteins from endoplasmic reticulum (ER) appear first in *cis*-cisternae, and then move on to medial and *trans*-cisternae before entering the *trans*-Golgi network, where they are sorted and targeted to their final destination, namely to lysosomes, the plasma membrane or the extracellular matrix (Kornfeld & Kornfeld 1985; Griffiths & Simons 1986; Hirschberg & Snider 1987). The many biochemical similarities in the processing of *N*-linked glycoproteins in plant and animal cells has led to the general assumption that the functional organisation of the enzymes involved in the synthesis of such glycoproteins is similar in plant and animal Golgi stacks (Kaushal *et al.* 1988; Faye *et al.* 1989; Jones & Robinson 1989). However, only in developing storage tissues of seeds is the GApp primarily involved in the modification of *N*-linked glycoproteins. In most non-seed tissues, the major products of the GApp are complex cell wall polysaccharides; glycoprotein processing appears to represent only a minor fraction of the overall Golgi activity (Robinson & Kristen 1982).

Complex cell wall polysaccharides such as the hemicellulose, xyloglucan (XG), and pectic polysaccharides, such as polygalacturonic acid

(PGA) and rhamnogalacturonan I (RG-I), appear to be both synthesised and processed in the GApp (Camirand & Maclachlan 1986; Camirand *et al.* 1987; Moore & Staehelin 1988). These cell wall matrix polysaccharides are very complex molecules, and neither their molecular structure nor their mode of synthesis is well understood (Bacic *et al.* 1988; Hayashi 1989). Furthermore, although the overall accuracy of the assembly process of complex polysaccharides is unknown (Waldron & Brett 1985), some polysaccharides have been shown to require a precise structure both for their structural interactions and their regulatory functions (Darvill & Albersheim 1984; Ryan 1987). For example, the growth regulatory activity of fragments of XG depends on the presence of both xylose and xylose-galactose-fucose side-chains in a specific spatial relationship (York *et al.* 1984; McDougall & Fry 1988). Two theoretical, non-exclusive mechanisms for producing error-free polysaccharides are (1) synthesis by extremely specific enzymes that can both distinguish different types of precursor molecules and catalyse the addition of specific sugars via defined linkages, or (2) segregation of enzymes involved in the synthesis of different polysaccharides into separate compartments of the GApp.

The central goal of our research is to sort out the spatial and functional relationship of the pathways for the synthesis, processing, sorting and packaging of glycoproteins and complex polysaccharides in the GApp of plant cells. Work to date has shown that Golgi stacks in different types of cell differ in their morphology, that a given Golgi stack can process simultaneously glycoproteins and polysaccharides, and that different polysaccharides are confined to different processing, sorting and packaging pathways.

Composition of the primary cell wall

Plant cell walls consist of cellulose microfibrils embedded in a matrix of polysaccharide and protein. Cellulose, the most abundant component of the cell wall, is synthesised at the cell surface by cellulose-synthesising complexes which deposit the cellulose fibrils while moving laterally through the plasma membrane, thereby encircling the cell and creating a microfibrillar cocoon (Delmer 1987). The two major types of matrix polysaccharide are hemicelluloses and pectic polysaccharides, both of which are synthesised in Golgi stacks and transported to the cell surface in secretory vesicles. To study the synthesis of these matrix molecules we have raised polyclonal antibodies in rabbits against XG, a hemicellulose, and PGA/RG-I, a pectic polysaccharide (Moore & Staehelin 1988).

XG possesses a backbone of β-1,4-linked D-glucose residues as in cellulose, which enables it to associate with the surface of cellulose micro-

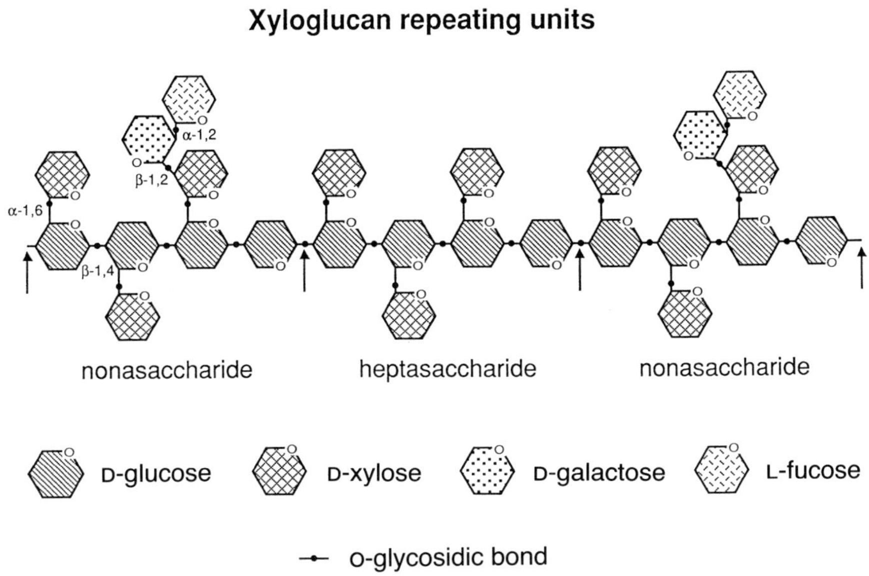

Fig. 1. Schematic model of the proposed partial structure for xyloglucan (XG). The heptasaccharide and nonasaccharide fragments are released upon endoglucanase treatment.

fibrils. Xylosyl side-chains are attached at O-6, and some of the xylosyl units are extended by the addition of D-galactose and L-fucose residues (Fig. 1; Hayashi 1989). Pea XG molecules have an average M_r of 300 000 and contain a backbone of about 1100 glucose residues representing a chain length of about 550 nm. Treatment of XG with the enzyme endoglucanase yields, among other fragments, a nona- and a hepta-saccharide in roughly equal amounts (Hayashi 1989). In functional terms, XG appears to limit self-association between cellulose microfibrils, control the extensibility of plant cell walls, and serve as a source for growth-regulatory oligosaccharide fragments (Ryan 1987; Hayashi 1989).

The molecules that comprise the PGA/RG-I fraction of pectic polysaccharides extracted from sycamore–maple suspension culture cells contain alternating domains of PGA and RG-I. In PGA, the galacturonic acid residues are connected by α-1,4 linkages, whereas the backbone of RG-I contains alternating 1,4-linked D-galacturonic acid residues and 1,2-linked L-rhamnose residues (McNeil *et al.* 1980, 1984; Selvendran 1985). The degree of polymerisation of RG-I is about 2000, with one-third of the residues making up the backbone. The poorly characterised side-chains are composed of arabinose, galactose and small amounts of fucose, and

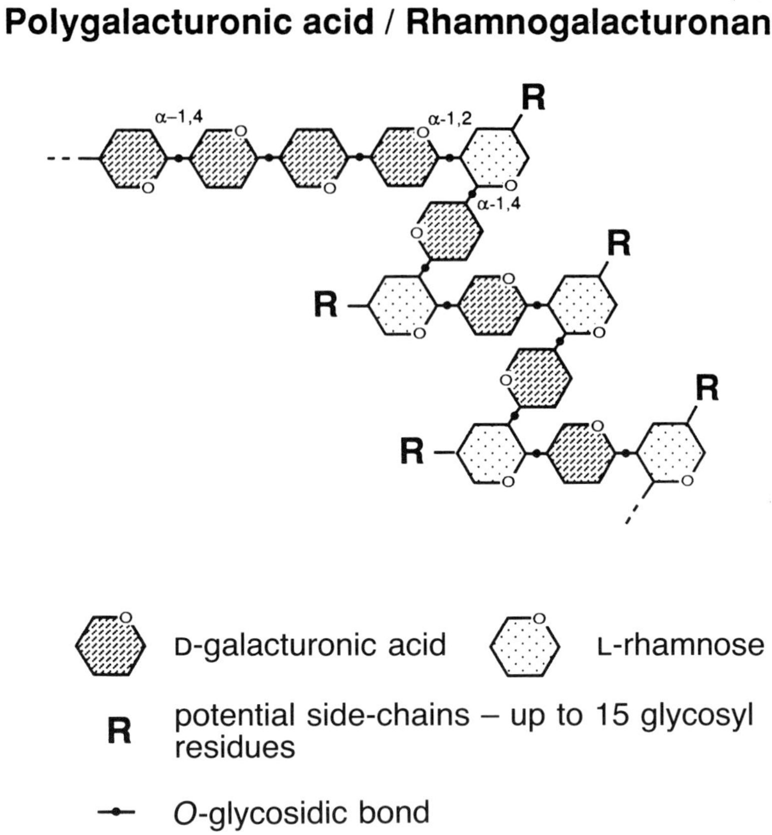

Fig. 2. Schematic model of the proposed structure for a segment of a mixed polygalacturonic acid/rhamnogalacturanan I (PGA/RG-I) polymer. The side-chains are composed of arabinose, galactose and small amounts of fucose and are attached through O-4 linkages to some of the rhamnose residues.

are attached to a fraction of the rhamnose units of the backbone (Fig. 2). Up to 70% of the PGA residues are esterified at the time of secretion, and become de-esterified by cell wall localised enzymes during cell wall maturation. This de-esterification enables the PGA domains to form characteristic calcium pectate gels. Pectic polysaccharides have been postulated to serve as 'glue' molecules between adjacent cell walls, and have been shown to be the source of oligosaccharide fragments that serve regulatory functions, including the induction of the hypersensitive cell death response (Darvill & Albersheim 1984; Ryan 1987).

A third cell wall matrix component to which we have obtained polyclonal antibodies is extensin, a hydroxyproline-rich, structural glycoprotein found in the cell walls of dicotyledonous plants (Lamport & Catt 1981). The rod-shaped extensin monomers are secreted into the wall and 'immobilised' there, presumably by the formation of covalent isodityrosine crosslinks between molecules (Fry 1988). The exact functions of extensin are still unclear, but, in general, their incorporation into cell walls has been correlated with an increase in mechanical strength, thereby enhancing resistance to a variety of stresses and plant pathogens (Swords & Staehelin 1989).

Localisation of matrix molecules in cell walls

We have used our anti-XG, anti-PGA/RG-I and anti-extensin antibodies (Moore & Staehelin 1988; Stafstrom & Staehelin 1988) in conjunction with immunocytochemical methods to localise these molecules in primary cell walls of clover root tip and carrot taproot tissues. Primary walls consist of three layers: two cellulose microfibril-containing layers that border the plasma membranes of the adjacent cell, and a middle lamella between the two cellulose-containing layers. The middle lamella is often tenuously thin between the two cellulose-containing layers, but is greatly expanded where three cells adjoin. Middle lamella material also appears to line the air spaces.

All three antibodies used in our studies label either the cell wall proper or the middle lamella material, but not both cell wall domains. Thus, anti-XG antibodies bind nearly exclusively to the cell wall proper, consistent with the hypothesis that XG is associated with cellulose fibrils. In contrast the anti-PGA/RG-I antibodies (note: raised against *de*-esterified PGA/RG-I) label only the middle lamella, and particularly the expanded middle lamella regions at the corner junctions between cells (Moore & Staehelin 1988). Interestingly, even though extensin is a highly basic molecule and might be expected to be strongly attracted to the acidic middle lamella regions, our anti-extensin antibodies show it to be localised only in the cell wall proper. Indeed, the pectic polysaccharides in the middle lamella appear to exclude extensin molecules (Stafstrom & Staehelin 1988).

Structural polarity and variability of plant Golgi stacks

Morphology of high pressure frozen root cap cells

Root cap cells are derived from proximal cells of the root apical meristem. During development they first expand and differentiate into

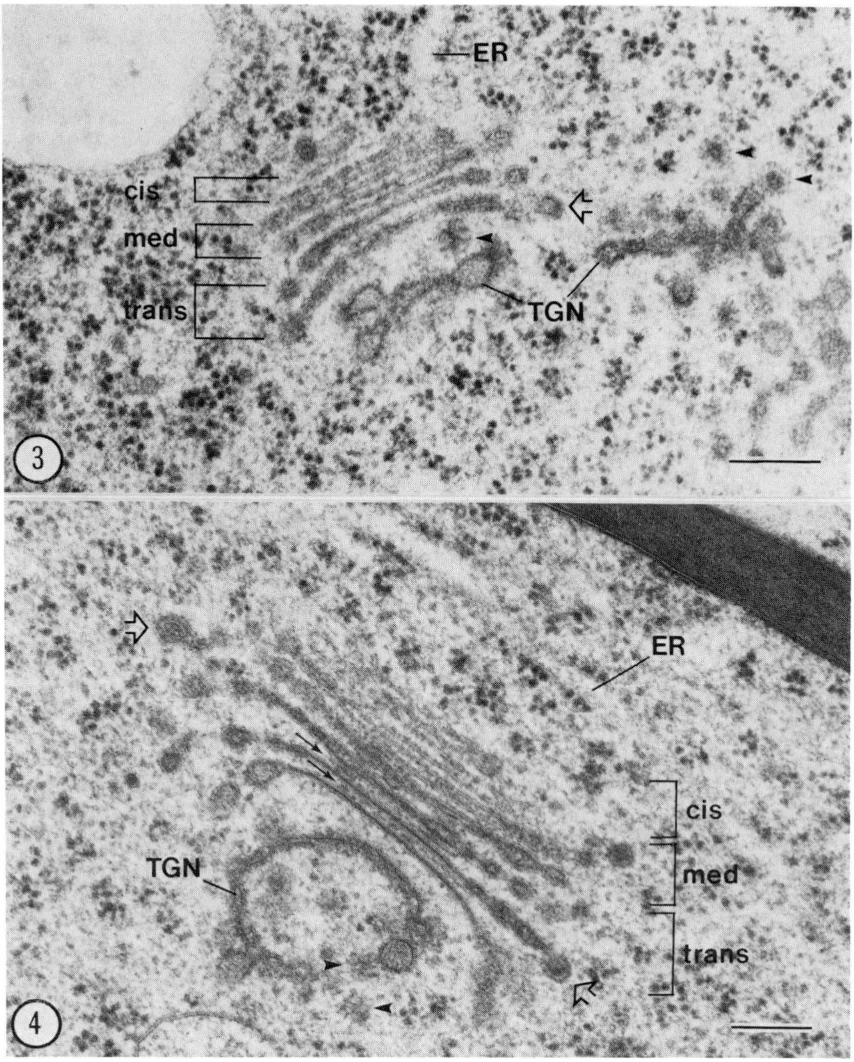

Figs. 3 and 4. (3) Golgi stack and *trans*-Golgi network (TGN) in a meristematic cell of a *Nicotiana* root tip. *Cis*-, medial and *trans* types of cisterna can be distinguished. The open arrow points to a non-clathrin-coated budding vesicle, the arrowheads to clathrin-coated vesicles. ER, endoplasmic reticulum. The bar represents 0.2 μm. (4) Golgi stack and associated *trans*-Golgi network (TGN) in a columella cell of a *Nicotiana* root cap. Differences in morphology of *cis*-, medial and *trans*-cisternae are more pronounced than for the meristematic Golgi stacks (Fig. 3). Intercisternal elements (small arrow) are seen at the medial to *trans*-

columella cells whose starch-rich amyloplasts serve as statoliths in the gravitropic response (Kiss *et al.* 1989). Subsequently, as the columella cells are displaced towards the root cap surface, they are converted to peripheral cells, the primary function of which is to produce slime to protect the growing root tip from mechanical damage by soil particles and from attacks by pathogenic bacteria and fungi (Sack & Kiss 1989).

We have employed high-pressure freezing (Moore 1987; Dahl & Staehelin 1989) in conjunction with freeze-substitution methods to increase the quality of ultrastructural preservation of *Arabidopsis* and *Nicotiana* root cap cells, and thereby obtain more precise information on structural parameters of Golgi stacks in such cells. As demonstrated in Figs. 3 to 5 the Golgi stacks of cells prepared in this manner reveal a degree of structural preservation that simply cannot be achieved with chemically fixed specimens. More importantly the micrographs reveal distinct differences in the morphology of Golgi stacks of meristematic, columnar, and young and old peripheral cells that may be related to differences in the synthetic programmes of the Golgi stacks in these cell types.

Golgi stacks of meristematic cells

Meristematic cell Golgi stacks (Fig. 3) tend to be relatively small (0.5–0.7 μm in diameter) and compact and possess five to six cisternae and an associated *trans*-Golgi network. Both the cisternae and the *trans*-Golgi network are enclosed in a fine filamentous matrix that defines the zone of exclusion of ribosomes. Each Golgi stack exhibits a distinct general *cis*-to-*trans* polarity in terms of width and staining of lumenal contents, and, upon close examination, three subtypes of cisternae (two each of *cis*, medial and *trans*) can be distinguished. *Cis*-cisternae have the most irregular contours, the widest lumens, and their lumenal contents pick up little stain. The membranes of medial cisternae tend to be straighter, enclose a narrower lumen, and exhibit discrete staining on the lumenal side of the membranes. In *trans*-cisternae the membranes often appear collapsed upon themselves, obliterating most of the lumen in the central region of the cisternae. However, overall the *trans*-cisternal membranes of meristematic cell Golgi stacks (Fig. 3) appear to be less collapsed than those of columella (Fig. 4) and of young peripheral cells (not shown). The bulging margins of the cisternae contain concentrated products, and some

---

Figs. 3 and 4 (*contd*)

cisternal interface and between *trans*-cisternae. Other labels as for Fig. 3. The bar represents 0.2 μm. (Figs. 3 and 4 from Staehelin *et al.* 1990, with permission.)

of the budding vesicles are covered by a non-clathrin type of coat (Orci *et al.* 1986). In contrast, clathrin-coated vesicles are seen associated with elements of the *trans*-Golgi network.

Golgi stacks of columella cells

Golgi stacks of columella cells (Fig. 4) are close to twice as wide (~ 1.2 μm) as those of meristematic cells, and are usually composed of seven cisternae (two *cis*, two medial, three *trans*). The staining pattern for the *cis*, medial and *trans*-cisternae are similar to those described for meristematic Golgi stacks, but the differences are more pronounced. The lumenal contents of the ~ 30 μm wide *cis* cisternae stain only very lightly. The medial cisternae are narrower (~ 20 μm) and are filled with more darkly staining products that are concentrated in the budding peripheral vesicles. Intercisternal elements (Mollenhauer 1965) appear at the transition to the *trans* cisternae, whose lumenal surfaces are tightly appressed (4–6 nm wide) in the central domains. Non-clathrin-coated vesicles are associated with the margins of some of the cisternae, whereas clathrin-coated vesicles are seen exclusively on vesicles of the *trans*-Golgi networks. The zone of exclusion of ribosomes stands out less clearly than in meristematic cells, due to the overall lower density of ribosomes in the cytoplasm.

Golgi stacks of peripheral cells

The main function of peripheral cells appears to be the secretion of root cap slime (Rougier 1981). Young peripheral cells still contain sizeable numbers of starch-filled amyloplasts and their Golgi stacks appear to be similar to those of the columella cells from which they are derived. Old peripheral cells are devoid of starch-filled amyloplasts and possess Golgi stacks with typical hypertrophied cisternal margins (Fig. 5).

During this developmental progression, several changes in Golgi morphology are seen: an increase of *trans*-cisternae by one or two units, the accumulation of darkly staining products in the *trans*-cisternae, the progressive loss in definition of *cis*- and medial cisternae, and the disappearance of the *trans*-Golgi network by the time the *trans*-cisternae become hypertrophied. Intercisternal elements (Mollenhauer 1965) are associated with all of the slime-producing *trans*-Golgi cisternae. Freeze-fracture micrographs of ultrarapidly frozen, slime-producing Golgi stacks demonstrate rows of intramembrane particles in *trans*-cisternal membranes (not shown). The spacing of these particle rows coincides with the spacing of the intercisternal elements.

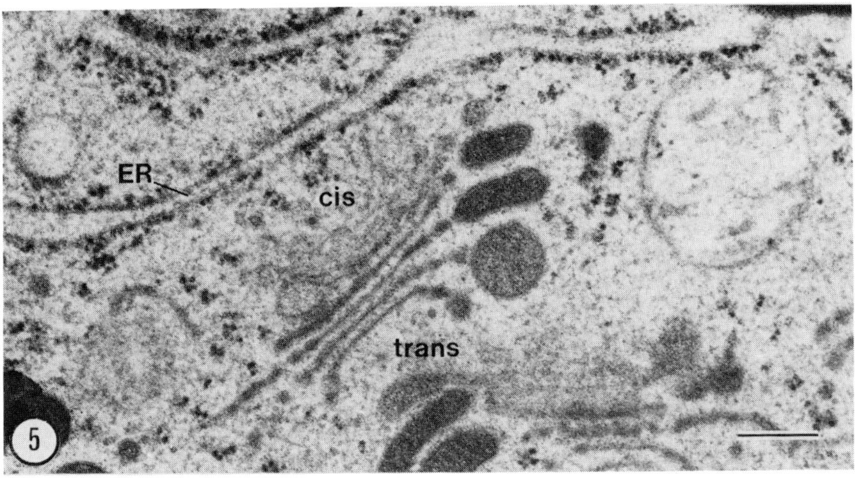

Fig. 5. Golgi stack in an older peripheral cell of an *Arabidopsis* root cap. The dark staining of the *trans*-cisternae contrasts with the very light staining of the *cis*- and medial cisternae. Note the absence of a *trans*-Golgi network. The bar represents 0.2 μm. (From Staehelin *et al.* 1990, with permission.)

Functional significance of the tissue-specific differences in Golgi architecture

One of the most striking aspects of this study is the extent of change of Golgi stack architecture during the developmental progression of root cap cells. Nevertheless, one should not overlook the fact that, despite these changes, the basic morphology of all plant and animal stacks is remarkably constant. Thus one can speculate that this basic organisation is dictated by common functional parameters, and that the variable features represent adaptations to functional requirements unique for a given type of tissue. To this end, the morphological distinction of *cis*, medial and *trans* subtypes of cisternae and the organisation of the *trans*-Golgi network probably relates to the sequential processing, sorting and packaging of glycoproteins that is common to all animal and plant Golgi stacks (Kornfeld & Kornfeld 1985; Hirschberg & Snider 1987; Faye *et al.* 1989; Jones & Robinson 1989). On the other hand, the unique structural features of plant Golgi stacks, such as the intercisternal elements and the tightly collapsed *trans*-cisternae, are most likely related to functions that are unique to Golgi stacks of plant cells, particularly the synthesis of polysaccharides. Furthermore, the variations in architecture of Golgi stacks in different types of root cap cell may reflect subtle shifts in synthetic activities, such as changes in glycoprotein to polysaccharide ratios, and shifts in ratios and types of polysaccharides that are produced.

Immunocytochemical localisation of glycoproteins and polysaccharides in plant Golgi stacks

Localisation of β-1,3-glucanase, extensin, XG and PGA/RG-I in Golgi stacks

There is a general consensus that glycoproteins destined for export, or for transfer to lysosomes or vacuoles, are synthesised in the ER, and then sequentially transferred to *cis*-, medial and *trans*-cisternae for processing before being sorted and packaged in the *trans*-Golgi network (Farquhar 1985; Griffiths & Simons 1986; Hirschberg & Snider 1987). This model predicts that antibodies raised against appropriate domains of a secretory glycoprotein should be able to immunolabel all cisternae of a Golgi stack actively involved in the processing of this glycoprotein. This has been found to be the case for the ethylene-induced defence protein, β-1,3-glucanase (Mauch & Staehelin 1989), and the cell wall glycoprotein extensin (Fig. 6).

As demonstrated by Moore & Staehelin (1988), polyclonal rabbit antibodies raised against two complex polysaccharides of plant cell walls, the hemicellulose XG, and the pectic polysaccharide PGA/RG-I, can also be used to immunolocalise these polysaccharides to Golgi stacks (Figs. 6 to 8). With these probes in hand, one can address the question of whether a given plant Golgi stack can process simultaneously both glycoproteins and polysaccharides, or if these two types of molecule are synthesised by different Golgi stacks. Figure 6 demonstrates that a given Golgi stack from a carrot suspension-cultured cell can simultaneously process the cell wall glycoprotein extensin and the polysaccharide XG.

Upon close examination of the Golgi stack, labelling patterns of the anti-PGA/RG-I and the anti-XG antibodies (Figs. 7 and 8), it may be seen that the distribution of the two labels is non-uniform across the stacks. This non-random labelling has been statistically verified to demonstrate that the anti-PGA/RG-I binding occurs nearly exclusively over *cis*- and medial cisternae, and the anti-XG labelling over *trans*-cisternae and the *trans*-Golgi network. These findings can be interpreted in different ways. For instance, our anti-XG antibodies may not recognise precursor forms of XG, and therefore not label *cis*- and medial cisternae. Similarly, our anti-PGA/RG-I antibodies may not bind tightly to mature (esterified?) forms of these pectic polysaccharides, and therefore not label *trans*-cisternae and the *trans*-Golgi network. Alternatively, these

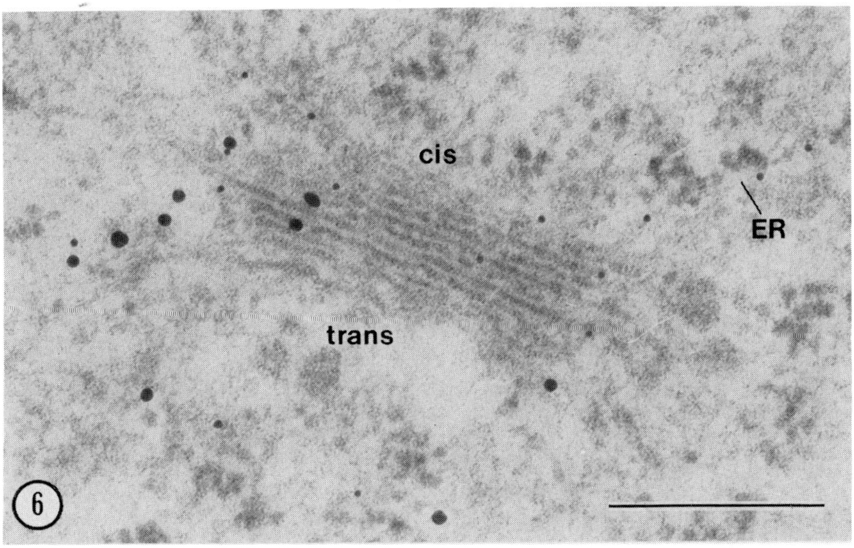

Fig. 6. Golgi stack of a carrot suspension-cultured cell double labelled with anti-extensin (7 nm gold) and anti XG (17 nm gold) antibodies. Note that the same Golgi stack labels with both antibodies with the extensin labelling extending across the entire stack and the XG labelling being confined to the *trans* side of the stack and the *trans*-Golgi network. The bar represents 0.2 μm. ER, endoplasmic reticulum.

observations are also consistent with the hypothesis (Fig. 9) that PGA/RG-I could be synthesised in *cis*- and medial cisternae only and also exit from medial cisternae, and that XG is made exclusively in *trans*-cisternae and leaves from the *trans*-Golgi network. Our current research is devoted to testing these hypotheses to learn more about the synthesis/packaging/secretion pathways of complex cell wall polysaccharides.

Tests of a new model of synthesis and secretion of complex plant polysaccharides

The model of the synthesis/secretion pathway for the complex, cell wall polysaccharides XG and PGA/RG-I shown in Fig. 9 enables several predictions to be made, some of which we have tested. For example, the model predicts that XG and PGA/RG-I should be packaged into separate secretory vesicles. This prediction can be tested using double labelling experiments in which the two antibodies are sequentially applied and labelled with protein A-gold particles of different sizes. In our study of

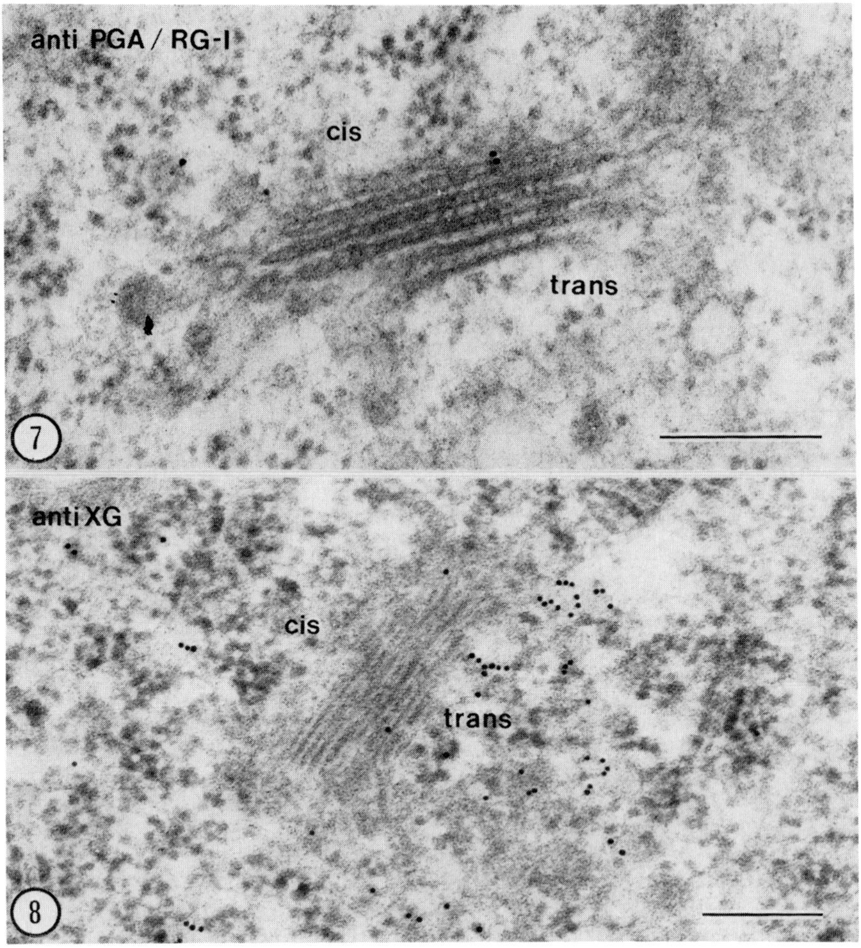

Figs. 7 and 8. (7) Anti-PGA/RG-I-labelled clover root tip Golgi stack. Labelling is limited to the *cis* side of the stack. The bar represents 0.2 μm. (8) Anti-XG-labelled clover root tip Golgi stack. Note the nearly exclusive labelling of *trans*-cisternae and the *trans*-Golgi network. The bar represents 0.2 μm.

clover root tip cells, 48% of the vesicles marked by two or more gold particles labelled only for XG, 43% only for PGA/RG-I, and 9% exhibited a mixed labelling pattern.

A second prediction of our model is that when secretion is blocked on the *trans* side of the Golgi stacks, for instance with the ionophore monensin (Morré *et al.* 1983), XG but not PGA/RG-I should accumulate

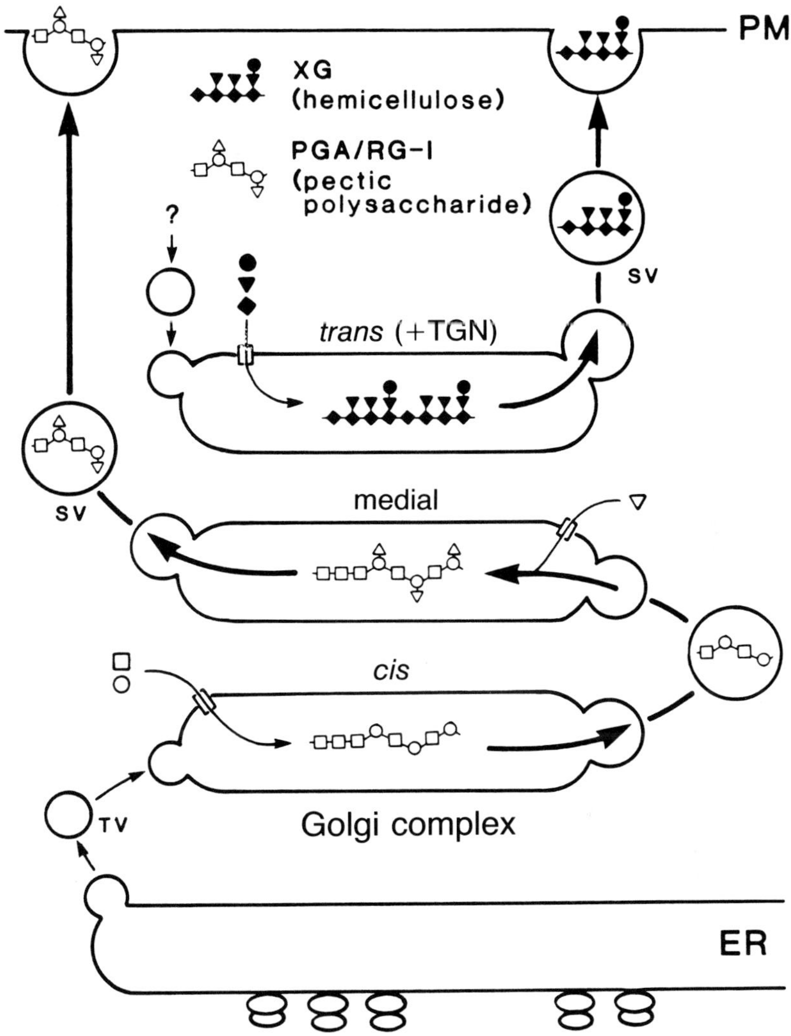

Fig. 9. Model for the synthesis of complex polysaccharides in the Golgi apparatus of plant cells. Synthetic enzymes for different classes of polysaccharide are postulated to be segregated to different Golgi cisternae. Thus, *cis*- and medial cisternae are shown to be associated with the synthesis of PGA/RG-I and *trans*-cisternae (and the *trans*-Golgi network) with the synthesis of XG. The model also depicts PGA/RG-I leaving the stack in vesicles derived from medial cisternae, and XG being packaged into separate vesicles on the *trans* side of the stack. PM, plasma membrane; sv, secretory vesicle; tv, transport vesicle; ER, endoplasmic reticulum.

in the blocked cisternae. This has been found to be the case (data not shown).

A third prediction is that if sorting/packaging of the cell wall glycoprotein, extensin, occurs in the same *trans*-Golgi network compartment as XG, then both extensin and XG should accumulate in the same compartment, when secretion is blocked by monensin. This too has been demonstrated to be the case (data not shown).

Conclusion

The evidence cited above is consistent with the hypothesis of complex polysaccharide synthesis outlined in Fig. 9, in which different types of polysaccharide are synthesised in, and secreted from, different Golgi cisternae. However, a number of critical experiments, such as determining biochemically that monensin inhibits the secretion of XG but not that of PGA/RG-I, need to be carried out before this highly speculative hypothesis should be viewed as an established model of polysaccharide synthesis.

Acknowledgements

This work has been supported by NSF grant DCB 8615763 and NIH grant GM-18639. Thanks are due to Janet Meehl for her help in preparing the Figures.

References

Bacic, A., Harris, P.J. & Stone, B.A. (1988). Structure and function of plant cell walls. In *The Biochemistry of Plants*, vol. 14, ed. P.K. Stumpf & E.E. Conn, pp. 297–371. San Diego: Academic Press.

Camirand, A., Brummel, D. & Maclachlan, G. (1987). Fucosylation of xyloglucan: localization of the transferase in dictyosomes of pea stem cells. *Plant Physiology*, **84**, 753–6.

Camirand, A. & Maclachlan, G. (1986). Biosynthesis of the fucose-containing xyloglucan nonasaccharide by pea microsomal membranes. *Plant Physiology*, **82**, 379–83.

Dahl, R. & Staehelin, L.A. (1989). High pressure freezing for the preservation of biological structure: theory and practice. *Journal of Electron Microscopy Technique*, **13**, 165–74.

Darvill, A.G. & Albersheim, P. (1984). Phytoalexins and their elicitors – a defense against microbial infection in plants. *Annual Review of Plant Physiology*, **35**, 243–75.

Delmer, D.P. (1987). Cellulose biosynthesis. *Annual Review of Plant Physiology*, **38**, 259–90.

Farquhar, A.G. (1985). Progress in unravelling pathways of Golgi traffic. *Annual Review of Cell Biology*, **1**, 447–88.

Faye, L., Johnson, K.D., Stern, A. & Chrispeels, M.J. (1989). Structure, biosynthesis, and function of asparagine-linked glycans on plant glycoproteins. *Physiologia Plantarum*, **75**, 309–14.

Fry, S.C. (1988). *The Growing Plant Cell Wall: Chemical and Metabolic Analysis*. Harlow, Essex: Longman Scientific & Technical.

Griffiths, G. & Simons, K. (1986). The trans Golgi network: sorting at the exit site of the Golgi complex. *Science*, **234**, 438–43.

Hayashi, T. (1989). Xyloglucans in the primary cell wall. *Annual Review of Plant Physiology and Molecular Biology*, **40**, 139–68.

Hirschberg, C.B. & Snider, M.D. (1987). Topography of glycosylation in the rough endoplasmic reticulum and Golgi apparatus. *Annual Review of Biochemistry*, **56**, 63–87.

Jones, R. & Robinson, D.G. (1989). Protein secretion in plants. *New Phytologist*, **111**, 567–98.

Kaushal, G.P., Szumilo, T. & Elbein, A.D. (1988). Structure and biosynthesis of plant *N*-linked glycoproteins. In *The Biochemistry of Plants*, vol. 14, ed. J. Preiss, pp. 421–64. New York: Academic Press.

Kiss, J.Z., Hertel, R. & Sack, F.D. (1989). Amyloplasts are necessary for full gravitropic sensitivity in roots of *Arabidopsis thaliana*. *Planta*, **177**, 198–206.

Kornfeld, R. & Kornfeld, S. (1985). Assembly of asparagine-linked oligosaccharides. *Annual Review of Biochemistry*, **54**, 631–64.

Lamport, D.T.A. & Catt, J.W. (1981). Glycoproteins and enzymes of the cell wall. In *Encyclopedia of Plant Physiology, New Series*, vol. 13B, ed. W. Tanner & F. Loewers, pp. 133–65. Heidelberg: Springer-Verlag.

Mauch, F. & Staehelin, L.A. (1989). Functional implications of the subcellular localisation of ethylene-induced chitinase and β-1,3 glucanse in bean leaves. *Plant Cell*, **1**, 447–57.

McDougall, G.J. & Fry, S.C. (1988). Inhibition of auxin-stimulated growth of pea stem segments by a specific nonasaccharide of xyloglucan. *Planta*, **175**, 412–16.

McNeil, M., Darvill, A.G. & Albersheim, P. (1980). The structure of plant cell walls. X. Rhamnogalacturonan I, a structurally complex polysaccharide in the walls of suspension-cultured sycamore cells. *Plant Physiology*, **66**, 1128–34.

McNeil, M., Darvill, A.G., Fry, S.C. & Albersheim, P. (1984). Structure and function of the primary cell walls of plants. *Annual Review of Biochemistry*, **8**, 625–63.

Mollenhauer, H.H. (1965). An intercisternal structure in the Golgi apparatus. *Journal of Cell Biology*, **24**, 504–11.

Moore, P.J. (1987). Location, synthesis and secretion of the cell wall matrix polysaccharides rhamnogalacturonan I and xyloglucan. Ph.D. thesis, University of Colorado.

Moore, P.J. & Staehelin, L.A. (1988). Immunogold localization of the cell-wall matrix polysaccharides rhamnogalacturonan I and xyloglucan during cell expansion and cytokinesis in *Trifolium pratense* L.: implications for secretory pathways. *Planta*, **174**, 433–45.

Morré, D.J., Boss, W.F., Grimes, H. & Mollenhauer, H. H. (1983). Kinetics of Golgi apparatus membrane flux following monensin treatment of embryonic carrot cells. *European Journal of Cell Biology*, **30**, 25–32.

Orci, L., Glick, B. & Rothman, J.E. (1986). A new type of coated vesicular carrier that appears not to contain clathrin: its possible role in protein transport within the Golgi stack. *Cell*, **46**, 171–84.

Robinson, D.G. & Kristen, U. (1982). Membrane flow via the Golgi apparatus of higher plant cells. *International Review of Cytology*, **77**, 89–127.

Rougier, M. (1981). Secretory activity of the root cap. In *Encyclopedia of Plant Physiology, New Series*, vol. 13B, ed. W. Tanner & F. Loewers, pp. 542–74. Heidelberg: Springer-Verlag.

Ryan, C.A. (1987). Oligosaccharide signalling in plants. *Annual Review of Cell Biology*, **3**, 295–317.

Sack, F.D. & Kiss, J.Z. (1989). Root cap structure in wild type and in a starchless mutant of *Arabidopsis*. *American Journal of Botany*, **76**, 454–64.

Selvendran, R.R. (1985). Developments in the chemistry and biochemistry of pectic and hemicellulosic polymers. *Journal of Cell Science Supplement*, **2**, 51–88.

Stafstrom, J.P. & Staehelin, L.A. (1988). Antibody localization of extensin in cell walls of carrot storage root. *Planta*, **174**, 321–32.

Staehelin, L.A., Giddings, T.H., Kiss, J.Z. & Sack, F.D. (1990). Macromolecular differentiation of Golgi stacks in root tips of *Arabidopsis* and *Nicotiana* seedlings as visualized in high pressure frozen and freeze-substituted samples. *Protoplasma*, **157**, 75–91.

Swords, K.M.M. & Staehelin, L.A. (1989). Analysis of extensin structure in plant cell walls. In *Modern Methods of Plant Analysis, New Series*, vol. 10, ed. H.F. Linskens & J.F. Jackson, pp. 219–33. Heidelberg: Springer-Verlag.

Waldron, K.K. & Brett, C.T. (1985). Interaction of enzymes involved in cell-wall heteropolysaccharide biosynthesis. In *Biochemistry of Plant Cell Walls*, ed. C.T. Brett & J.R. Hillman, pp. 79–98. Cambridge: Cambridge University Press.

York, W.S., Darvill, A.G. & Albersheim, P. (1984). Inhibition of 2,4-dichlorophenoxyacetic acid-stimulated elongation of pea stem segments by a xyloglucan oligosaccharide. *Plant Physiology*, **75**, 295–7.

E. KRONESTEDT-ROBARDS AND
A.W. ROBARDS

Exocytosis in gland cells

Glands

Plant glands consist of highly specialised cells in well-defined arrangements which release specific substances (secretory products). Glandular activity is of considerable importance for plants with respect to their physical and biological environments: e.g. in maintaining the salt balance at an optimal level (in halophytes), in attracting pollinators or in deterring herbivores and microorganisms.

In many cases glandular cells discharge extremely large amounts of secretory product. The high metabolic rate of such cells makes them exciting objects for studies on intra- (and inter-) cellular processes. Substances supplied by the vascular tissues are usually only slightly modified when they are secreted, whereas other substances, such as protein and lipophilic materials, may be synthesised within the cells. Actively secreting cells have dense protoplasts, resulting from decreasing vacuolation during the maturation process, which makes them more amenable to ultrastructural investigations. Because of the high energy demands, secretory tissues usually contain numerous mitochondria. The frequency of other organelles varies greatly in relation to the material secreted (Fahn 1987). Over the last 30 years numerous different glands have been studied ultrastructurally but only a few of these investigations have combined cytological with physiological methods, which leaves too much for subjective interpretation and speculation. This means that information on the mechanism whereby the secretion leaves the gland cells is very limited despite the fact that there exists a vast literature on the subject. In the present chapter we review this literature briefly; we attempt to comment on those publications that contribute to our understanding of secretory mechanisms, particularly those relating to exocytotic processes, and we present some data from our own results, mainly on *Abutilon*.

Definitions

Before moving on, it may be useful to make some terminological clarifications.

Secretion

Secretion is the elimination of any substances which are transported to the exterior of the protoplast and not stored for remobilisation but withdrawn from further metabolic processes (Fahn 1979*b*; Schnepf 1974).

Two main types of secretion can be distinguished on the basis of the origin and location of the gland (Fahn 1979*b*, 1988*a,b*). In *exogenous* secretion the secretory tissues occur on the plant surfaces. The secreted substance is usually exuded directly to the exterior of the plant. In *endogenous* secretion the secretory tissues occur inside the plant body and secrete into specialised intracellular spaces. In some cases, as for instance in laticifers, secreted inert material is deposited inside cells; these can then be regarded as storage cells rather than gland cells, since there is no extrusion process. Structures of this kind are not discussed in this chapter, nor is the secretory process involving cell wall formation (as this is typical of all maturing plant cells), nor is the production of mucilage from the root cap.

The elimination of secretory products can take place in various ways. When they are discharged from an intact cytoplasm (merocrine secretion), they can pass through the plasmalemma directly (active process or concentration gradient – eccrine secretion), in which case transport occurs in molecular or ionic form through the plasmalemma (Fig. 1). Active transport involves a carrier system, which generally uses phosphorylation/dephosphorylation reactions (Eriksson 1977). An alternative method for the discharge of secretory products is through membrane-bounded secretion vesicles, which extrude their contents by exocytosis (granulocrine secretion) (Fig. 1). This type of secretion is also called vesicle extrusion or reverse pinocytosis. In the literature, granulocrine secretion has often been shown to be associated with acid polysaccharides secretion (see e.g. Schnepf 1969). Holocrine secretion does not involve a discharging process but the secretory cells degenerate and lyse. This may start as a merocrine secretion.

Exocytosis

The definition of exocytosis that we use is 'the discharge from a cell of molecules or particles that cannot diffuse or be directly transported through the plasmalemma'. Exocytosis can be stimulation dependent or independent (Allison & Davies 1974). It is usually envisaged that the

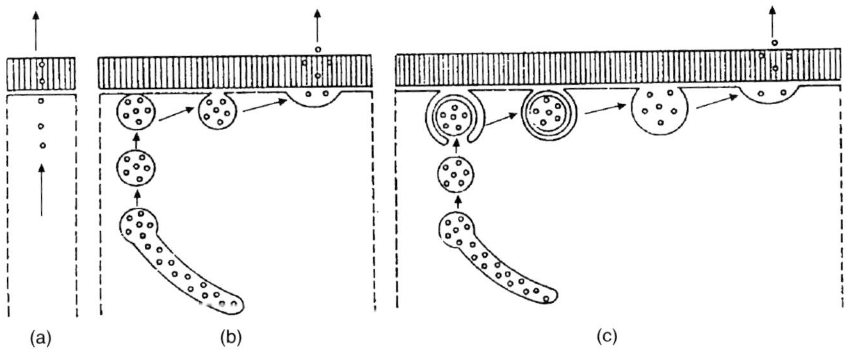

Fig. 1. The different pathways of secretion from secretory cells in vascular plants. (a) Membrane transport – *eccrine secretion*; (b) the membrane fuses with the plasmalemma – *granulocrine secretion*; (c) vesicle elimination is associated with invagination of the plasmalemma – *granulocrine secretion*. (Redrawn from Fahn 1979*b*.)

membrane of a secretion vesicle, derived from the endoplasmic reticulum (ER) or dictyosomes (Golgi bodies), fuses with the plasmalemma and becomes incorporated with it, so that the secreted material is deposited outside the protoplasm in the extracytoplasmic space. A transformation of the vesicle membrane so that it acquires a molecular identity with the plasmalemma is a requirement for this kind of extrusion process. It is also important to recognise the implication for membrane recycling of any exocytotic mechanism in a non-expanding cell (see Steer & O'Driscoll, this volume). The terms exocytosis and granulocrine secretion are used synonymously in this chapter.

Wall protuberances

Many types of glandular cell have wall protuberances which result in an enlarged plasmalemma surface area. In this way secretion is facilitated either by an increased area for membrane-bound carrier mechanisms or for passive diffusion. Wall protuberances have been reported from a wide range of secretory structures such as nectaries (see e.g. Fahn & Rachmilewitz 1970), salt glands (Shimony & Fahn 1968), glandular hairs of carnivorous plants (Schnepf 1961*b*; Scala *et al.* 1968) and glandular trichomes in the inflorescence of *Chrysanthemum* (Vermeer & Peterson 1979). They are rare along the outer walls of epidermal cells of nectaries which are multicellular outgrowths (Davis *et al.* 1988 on *Vicia faba*). The enlargement of the surface of the plasmalemma and the presence of acid phosphatase close to it have been taken as supporting evidence for eccrine secretion (Heinrich 1975*a,b*). Wall protuberances are thus thought to aid eccrine secretion of individual molecules and are not often found in systems where secretion by vesicles (granulocrine secretion) has been suggested (Lüttge 1971; Lüttge & Schnepf 1976).

Previous studies

Fahn and Schnepf have been two major contributors to the literature on secretion. By 1949 Fahn had already published on the ecology of nectar secretion, and by 1952 on floral nectaries. Subsequently he has written a major book on secretion in plants (Fahn 1979*b*) as well as a range of papers on secretion in plants, especially from nectaries and salt glands.

Schnepf has also published many significant papers on the structure and function of secretory cells, and on the elimination process in particular. The list of plants and the various types of gland that he has investigated is long, comprising about 20 different genera.

In this section we review briefly the literature on extrusion mechanisms in different plant glands. The distinction between types of gland is made according to the substance secreted, even though this is often a mixture of materials which are, in the main, hydrophilic. The construction of the different types of gland is not dealt with here but morphological and anatomical details can be found in original papers or handbooks.

Mucilage glands

'Pure' mucilage glands

Mucilage glands mainly secrete polysaccharide polymers and occur in many different plant groups. The morphology of these glands varies (e.g. idioblasts, trichomes or ducts). Common for all mucilage secretion is the involvement and apparent high activity of the Golgi apparatus. Numerous active Golgi bodies have been observed in mucilage-secreting cells at the secretory stage. The presence of polysaccharides in the Golgi vesicles has been demonstrated by histochemical and other methods (see e.g. Rougier 1972). It appears that the Golgi bodies are involved in mucilage production and that the mucilage is eliminated from the cytoplasm by granulocrine secretion (Fahn 1987; Schnepf 1973; Trachtenberg & Fahn 1981). Some authors (e.g. Horner & Lersten 1968) have suggested that the ER is also indirectly involved in mucilage secretion.

Functional mucilage glands in higher plants have been reported in glandular hairs of *Mimulus tilingii*, squamulae (glandular scales) of young *Elodea*, *Potamogeton* shoots, and the inner glands of *Hibiscus esculentus*. All of these apparently secrete the mucopolysaccharidic slime via Golgi vesicles and all of them lack wall protuberances (see Table 1 for references). The rate of exocytosis in the glandular cells in *Mimulus* were measured and found to be high, but not extremely so, by Schnepf & Busch (1976). In *Elodea*, periodic acid/thiocarbohydrazide/silver-proteinate was used to visualise polysaccharides in the Golgi vesicles and in the walls (Rougier 1972).

Digestive glands

Digestive glands occur in carnivorous plants. These capture the prey with a sticky slime, secrete digestive enzymes and absorb the digested products. Many plants, such as *Drosophyllum lusitanicum*, have leaves with two types of gland: those which secrete the mucilage, and the digesting structures. The Golgi has been shown to play the most important role in the extrusion process. As a result of simultaneous physiological and cytological investigations, overwhelming evidence demonstrates that the vesicles are secretory and that elimination is granulocrine, in contrast to other examples in higher plants where it has only been assumed (Lüttge & Schnepf 1976). The criteria used to confirm the granulocrine process were the staining character of the content of the vesicles, cytochemical reactions, autoradiographic evidence, the membrane of the vesicles becoming increasingly similar in structure to the plasmalemma with which it eventually fuses, and the correlation of cytological and physiological activity.

The outer layers of both types of gland in *Drosophyllum* have numerous small wall protuberances. The ER is inconspicuous. The rate of secretion was determined by estimating the amount of fluid secreted per hour through a glandular surface area of 1.0 mm^2. The interrelationship between the amount of secreted fluid and the number of secretion vesicles was as follows. At 28 °C the secretion intensity was 1.3 mm^3 mm^{-2} min^{-1}. Below 1 mm^2 of surface area there was a total volume of secretion vesicles of 3.4 mm^3. To discharge 1.3 mm^3 of fluid per minute by a vesicle mass of 3.4 mm^3, a turnover rate of 3.4 : 1.3 = 2.6 min would be necessary (order of magnitude; see Schnepf 1974). Metabolic inhibitors and anaerobic conditions prevented secretion and a correlated lack of secretion vesicles could then be observed (Schnepf 1973). This demonstrates that energy deficiency inhibits the formation of the secretory product rather than its discharge (Schnepf 1974).

Maturation of the contents of the secretion vesicles is associated with modification of the vesicle membrane. This became evident in one fixation where the plasmalemma had split artificially. As a result, in cross-section it appeared as two lines separated by a gap of up to 20 nm. In these cells the membranes of the Golgi cisternae and of the adherent

Table 1. *Literature citations of granulocrine secretions*

Glands	ER	Dict.	Not speci- fied	Wall prot. present	Source
Mucilage					
Mimulus		x			Schnepf & Busch 1976
Elodea (squamulae)		x			Rougier 1972
Potamogeton		x			Rougier 1969, 1972
Psychotria		x			Horner & Lersten 1967, 1968
Hibiscus		x			Mollenhauer 1967
Digestive					
Drosophyllum		x		x	Schnepf 1961*a*, *b*, 1965*b*, 1968, 1974, 1976, Rougier 1972
Drosera		x		x	Ciobanu & Tacina 1973; Dexheimer 1976
Pinguicula		x		x	Schnepf 1963, 1974; Vassilyev & Muravnik 1988
Dionaea	x				Robins & Juniper 1980*a*, *b*
Prot. carb. trich.					
Pharbitis	x				Healey 1967; Unzelman & Healey 1974; Schnepf 1975
Isoetes (ligule)	x				Kristen *et al.* 1982
Veronica		x			Kristen & Lockhausen 1985
Colleters					
Psychotria		x			Horner & Lersten 1967, 1968
Rumex, Rheum		x			Schnepf 1968
Salt glands					
Avicennia		x			Shimony *et al.* 1973
Limonium			x	x	Ziegler & Lüttge 1966; Faraday & Thomson 1986
Tamarix			x		Thomson & Liu 1967; Thomson *et al.* 1969
Hydathodes					
Phaseolus			x	(x)	Perrin 1970
Nectaries, floral					
Vicia	x			x	Davis *et al.* 1988
Diplotaxis			x		Eymé & LeBlanc 1963; Evmé 1966

Table 1. (*contd*)

Glands	ER	Dict.	Not specified	Wall prot. present	Source
Helleborus			x		Eymé & LeBlanc 1963
Abutilon	x				Findlay & Mercer 1971*b*; Robards & Stark 1988
Hibiscus	x				Sawidis *et al.* 1989
Fritillaria	x				Present work
Lonicera	x			x	Fahn & Rachmilevitz 1970, 1975; Fahn 1979*a*, *b*
Aleurites	x				Belin-Depoux & Clair-Maczulajtys 1975
Tropaeolum	x	x			Fahn & Benouaiche 1979; Rachmilevitz & Fahn 1975
Musa	x	x			Fahn 1979*a*, *b*
Aloe		x		x	Schnepf & Pross 1976
Aloe			x	x	Heinrich 1975*a*, *b*
Gasteria		x		x	Schnepf & Pross 1976
Coryanthes			(x)		Schnepf *et al.* 1983
Bromeliaceae (including *Billberiga*)		x			Benner & Schnepf 1975; Schnepf & Benner 1978
Citrus	x				Rachmilevitz & Fahn 1973
Vinca	x				Rachmilevitz & Fahn 1973; Fahn 1979*b*
Asclepias	x	x		x	Schnepf & Christ 1980, Christ & Schnepf 1988
Gomphocarpus	x	x		x	Christ & Schnepf 1988
Nectaries, extrafloral					
Impatiens	x				Figier 1972
Plumeria	x				Mohan & Inamdar 1986
Acacia			x		Marginson *et al.* 1985*a*, *b*
Gossypium	x				Eleftheriou & Hall 1983*a*, *b*
Vicia			x		Wrischer 1962; Davis *et al.* 1988
Glandular trichomes					
Mentha			x		Amelunxen 1964
Stigma, placenta					
Aptenia	x				Kristen 1977
Petunia	x				Kroh 1967
Lycopersicum			x		Dumas *et al.* 1978

Table 1. *(contd)*

Glands	ER	Dict.	Not speci- fied	Wall prot. present	Source
Secale			x		Heslop-Harrison & Heslop-Harrison 1980
Hordeum			x		Heslop-Harrison & Heslop-Harrison 1980
Forsythia	x				Dumas 1973
Capiscum (placenta)	x				Zamski *et al.* 1987
Aptenia (placenta)		x			Kristen 1976
Osmophores					
Stanhopea			x		Stern *et al.* 1987; Curry *et al.* 1988/9
Restrepia	x				Pridgeon & Stern 1983

Notes: ER, endoplasmic reticulum; Dict., dictysomes; protr., protruberances; Prot. carb. trich., protein-carbohydrate secreting trichomes.

The table illustrates views on how the exocytosis is most likely to occur in the particular plant; in some cases the statement has been more vague than the table may indicate.

vesicles were of the usual thickness. The membranes of the detached vesicles, however, were similar to the plasmalemma (Schnepf 1965*b*).

The secretion was temperature and water dependent. After interruption of the water supply, secreted material accumulated within the extracytoplasmic space. A sufficiently high turgor pressure was therefore apparently necessary to enable secretion to proceed further (Schnepf 1965*b*).

The Golgi apparatus is thought also to participate in the secretion from the digestive glands of *Drosera capensis* and *Pinguicula vulgaris* (Table 1). Gilchrist & Juniper (1974) observed a form of stimulated 'exocytosis' in *Drosera* when it was fed with a protein source and speculated that this may be a feature of those specialised organs that rapidly take up large quantities of externally applied nutrient.

In the digestive glands of *Dionaea muscipula*, on the other hand, secretion was suggested to occur by direct fusion of ER to the plasmalemma or via vesicles and vacuoles derived from the ER. Robins & Juniper (1980*a,b*) suggested this following the use of autoradiography to track the pathway of secretion and reported that neither dictyosomes nor dictyosomal vesicles appeared to be involved.

Protein-carbohydrate secreting trichomes

Secretory trichomes from epidermal cells on the leaf primordia and stem of *Pharbitis nil* (Convolvulaceae) (see Table 1) secrete a protein-carbohydrate mucilage which covers the shoot apex and developing leaves, thus causing a difference in release mechanism in relation to the polysaccharide-secreting glands (cf. Kristen 1978). At the stage of secretion in *Pharbitis*, numerous protein and carbohydrate storage vesicles were observed. Distended profiles of rough endoplasmic reticulum (RER) were frequently in direct contact with the plasmalemma, thus apparently forming an open continuum between the extracytoplasmic space and the storage vesicles (Fig. 2). It was concluded (Unzelman & Healey 1974) that it was coated vesicles and RER which were active in secretory transport and considered likely that the materials packaged into the storage vesicles were transferred to the RER at poculiform ('cup-shaped') junctions and subsequently transported from the cell through transient RER plasmalemma fusion. However, according to Schnepf (1975) an artificial fusion of different membranes could have occurred. Also, in the ligule of *Isoetes*, polysaccharides and proteins appear to be released from the ER but, in this case, from tubular elements. Kristen *et al.* (1982) found no indication of the secretion from the Golgi apparatus as later described by Kristen & Lockhausen (1985) for the secretory trichomes of submerged and emerged leaves of *Veronica beccabunga* (Table 1).

Colleters

Colleters, where present, are thought to protect young differentiating leaves or their stipules from desiccation, as they secrete 'bud-glue'. They constitute multicellular glandular trichomes, sometimes with a stalk, and secrete a sticky substance containing carbohydrate mucilage and lipophilic substances or protein. In the dendroid colleters of *Psychotria bacteriophila* and in ochrea hairs of buds in *Rumex* and *Rheum* (see Table 1) dictyosomes have been shown apparently producing exocytotic vesicles.

Salt glands

Salt glands usually eliminate salt to the *outside* of the gland, but in some plants they secrete it into the central vacuole (e.g. the salt glands of *Atriplex*; Osmond *et al.* 1969). The distinction between hydathodes and salt glands is not clear, as the secretion from hydathodes often also contains some salts. Salt glands are typically found in halophytes.

The views on how salt is eliminated from the cytoplasm include both granulocrine and eccrine possibilities. Shimony *et al.* (1973) identified three types of vesicle in *Avicennia*, one of which was regarded as being

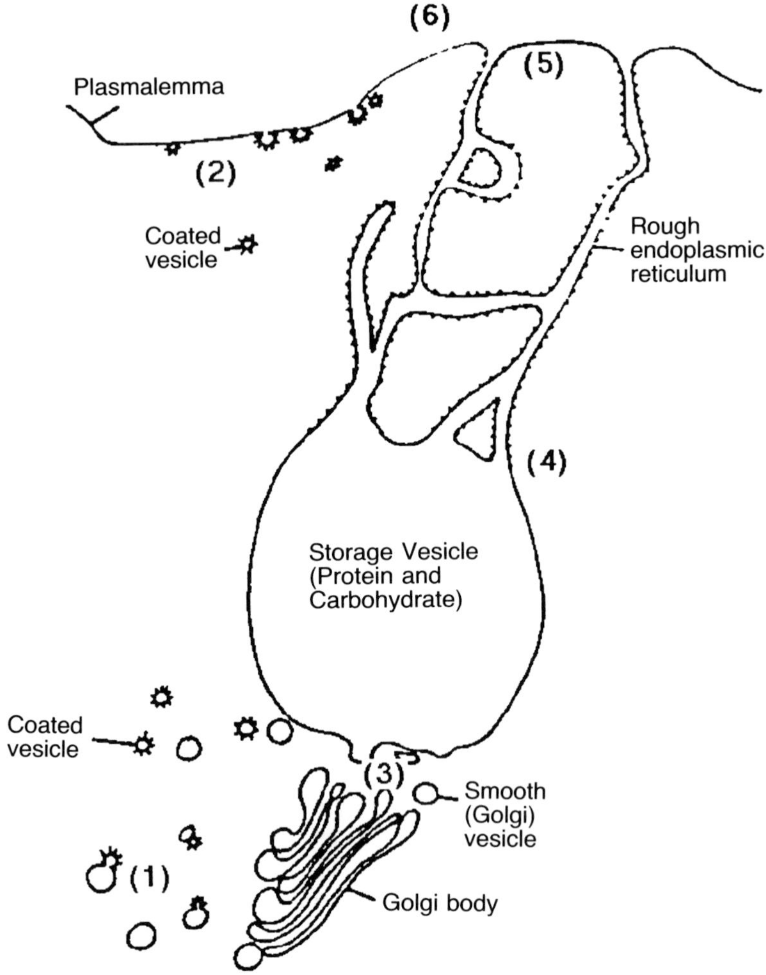

Fig. 2. Model of the manner of secretion of protein-carbohydrate mucilage by trichomes of *Pharbitis nil* as suggested by Unzelman & Healey (1974). 1, Golgi vesicles budding coated vesicles; 2, coated vesicles fuse with the plasmalemma; 3, smooth Golgi vesicles coalescing to form storage vesicle; 4, rough endoplasmic reticulum (RER) in direct contact with storage vesicle; 5, RER in direct contact with plasmalemma; 6, site of eliminated protein-carbohydrate mucilage. (Redrawn from Fahn 1979*b*.)

specifically involved in salt secretion. Vassilyev & Stepanova (1990) suggested that the wall protuberances in *Limonium* are rudimentary and presumably not involved directly in NaCl secretion, and they did not find any evidence for exocytosis. These authors presented a model in which the active step in NaCl secretion is the influx of ions into the vacuole; the efflux is believed to be passive and occurs from the vacuole into the periplasmic space through channels in the contracting area of tonoplast and plasmalemma.

The structure of the salt glands of *Tamarix* is similar to those of *Limonium* but less complex. In plants grown in a rubidium ion-containing medium the rubidium ions (as potassium analogues) were accumulated in 'micro-vacuoles', which were visible on electron micrographs, and which were subsequently secreted by fusion with the plasmalemma (Thomson *et al.* 1969). Hill & Hill (1973, 1976), however, opposed this on several physiological grounds. They suggested that the exudate in *Limonium* is produced by ionic pumping at the plasma membrane and that the vesicles are merely cross-sections of plasma membrane invaginations.

Hydathodes

Hydathodes exude water in liquid form, varying from almost pure water to weak inorganic and organic solutions. The process is called guttation and takes place under conditions of high soil moisture and low transpiration from the plant. The hydathode can be organised as a simple system of intercellular spaces or as a complex true gland with a resemblance to some nectaries. It therefore follows that the secretory mechanisms can be expected to vary between different types of hydathode. Generally, the secretion passes through the epidermis via modified stomata, the water-pores. The hydathodes are usually divided into those which are active or passive. Secretion from active hydathodes has been interpreted to be eccrine, as concluded from the amplification of the plasmalemma by simple invagination or by wall protuberances, and from the poor development of the Golgi apparatus. This has been shown in *Taraxacum officinale*, *Cichorium intybus* and *Papaver rhoeas* (Perrin 1971) and, apparently, in the trichome hydathodes of *Cicer* (Schnepf 1965*a*). In the trichome hydathodes of *Phaseolus multiflorus*, however, Perrin (1970) considered secretion to be exocytotic despite the presence of small wall protuberances.

Nectaries

Floral nectaries

The nectary is an organ that is developed especially to secrete sugars and to retain other substances (Lüttge 1961). The sugars are usually sucrose, glucose, fructose and various oligosaccharides; other constituents can be ninhydrin-positive substances, proteins, phosphates, organic acids, amino acids, mineral ions, vitamins, lipids and antioxidants.

Nectaries are found on the plant surfaces in the form of outgrowths, or deeply in the organ on which they occur (such as septal nectaries). Nectar is exuded by epidermal cells, often developed as trichomes, or by nectariferous parenchymatous cells which secrete into intercellular spaces and

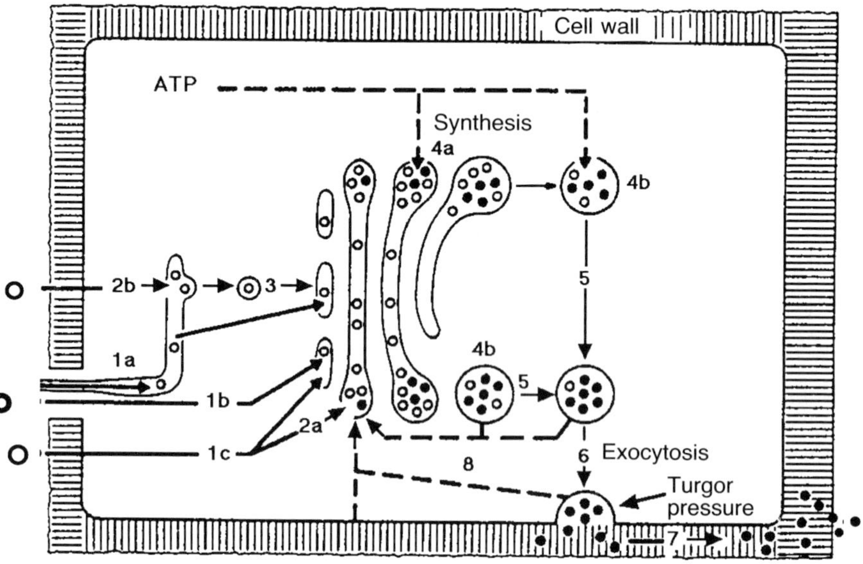

Fig. 3. Model of polysaccharide secretion via the Golgi apparatus. 1 a–c, Loading of the secretory cell through plasmodesmata or by membrane passage; 2 a,b, membrane passage into the Golgi or ER; 3, supply of the dictyosome by ER-derived vesicles with membrane material, synthesising enzymes and sugars (open circles); 4 a,b, polysaccharide (filled circles) synthesis in the dictyosomes and detached vesicles; 5, transport of secretion vesicles; 6, exocytosis; 7, transport and exudation of the secretion product into the cell wall by turgor pressure; 8, backflow of membrane material from the plasmalemma with the incorporated vesicle membranes and from secretion vesicles to the Golgi apparatus. (Redrawn from Lüttge & Schnepf 1976.)

Table 2. *Literature citations of glandular secretion regarding the release of secretory products*

Type of Gland ↓	No.[a]	Granulocrine ER	Granulocrine Dict.	Granulocrine ER *or* Dict.	Ecc-rine	Wall protr.	E/W[b]
Mainly hydrophilic secretion							
Mucilage glands	5		5				
Digestive glands	4	1	3			3	
Prot. carb. trich.	3	2	1				
Colleters	3		3				
Salt glands	16			2	14	2	1
Hydathodes	6			1	5	6	4
Nectary, floral	34	11	7	4	12	8	5
Nectaries, extrafloral	9	3	2		4		
Mainly lipophilic secretion							
Glandular trichomes	1			1			
Stigma	6	3		3			
Osmophores	4	1		1	2		

Notes: Wall protr., expresses presence of wall protuberances. ER, endoplasmic reticulum; Dict., dictyosome; Prot. carb. trich., protein-carbohydrate secreting trichomes.

[a]Number of investigated glands where an opinion has been expressed on how the secretory products leave the protoplast.

[b]Number of papers where eccrine secretion and wall protuberances have been combined.

from them via modified trichomes or stomata (Fahn 1979*a*). The diversity of nectaries correlates with a diversity of ultrastructural features, and to a diversity of ideas on secretory mechanisms. Lüttge & Schnepf (1976) proposed that, because of recognised diversity both in nectary type and nectar composition, both eccrine and granulocrine secretion may be possible (Fig. 3). Mechanisms for the release of nectar had also been discussed earlier by Ziegler (1965).

The majority of studies so far report the ER to be the most conspicuous feature in nectary cells at the time of secretion and the most likely organelle to give rise to exocytotic vesicles (see Tables 1 and 2). Heinrich (1975*a,b*) found high ATPase, nucleotide diphosphatase and glucose-6-phosphatase activity in the ER of sugar-secreting epithelial cells of *Aloe*.

This, in combination with the common absence of phosphatase activity in the plasma membrane, led him to suggest that the ER plays an important role in nectary sugar transport, again pointing to a granulocrine mode of prenectar secretion.

Dictyosome activity subsides with the onset of secretion in most species, although vesicles have also been reported from actively secreting nectaries (Table 1). Fahn (Fig. 4), together with Rachmilevitz (Rachmilevitz & Fahn 1975) and with Benouaiche (Fahn & Benouaiche 1979), has suggested the alternative possibility that both the ER and the Golgi apparatus participate in nectar secretion.

Many nectary cells are characterised by apical wall protuberances, especially the epithelial cells of nectaries within the gynoecium of several monocotyledons (e.g. Schnepf 1974). An eccrine mode of nectar secretion is usually suggested for nectaries with these wall ingrowths (Fahn 1987), such as in *Euphorbia pulcherrima* (Schnepf 1964, 1965*b*), in *Gasteria* (Schnepf 1964, 1974), in *Trifolium pratense* (Eriksson 1977), in *Aptenia cordifolia* (Meyberg & Kristen 1981), and in *Brassica napus* (Davis *et al.* 1986).

Strelitzia reginae has septal nectaries with wall protuberances (Kronestedt & Walles 1986). Kronestedt & Robards (1987) made a quantitative evaluation and Kronestedt-Robards *et al.* (1989) a qualitative analysis of the nectar. Transmembrane sugar fluxes were estimated to be in the range of 1.0×10^{-6} to 1.0×10^{-7} mol s^{-1} m^{-2}, which is among the highest transmembrane fluxes for sugars recorded in plants. This, among other reasons, led to the conclusion that eccrine secretion was the most likely possibility (Kronestedt & Robards 1987).

There are also septal nectaries where it has been considered that Golgi vesicles act as the secretory organelles (*Aloe* and *Gasteria* – Schnepf & Pross 1976), while there are others which have no true wall protuberances and an apparent granulocrine secretion (Bromeliaceae; Benner & Schnepf 1975; Schnepf & Benner 1978). Fiordi & Palandri (1982) suggested the possibility of both apoplastic and granulocrine pathways in *Tillandsia* (Bromeliaceae), some of the nectar constituents being drawn directly from the phloem.

Schnepf & Christ (1980; Christ & Schnepf 1988) discussed the question of whether secretion is eccrine or granulocrine in the nectaries of *Asclepias curassavica* and *A. tuberosa* and in *Gomphocarpus fruticosus* and again proposed the possibility that both mechanisms may be occurring simultaneously: the ER was swollen in active secretory cells but not in inactive or inhibited ones; vesicles in the periphery of the dictyosomes and in the *trans*-face showed a positive reaction for the Thiéry-reaction, which demonstrates the presence of polysaccharides; plasmalemma infoldings increased in size and number during secretion but were not

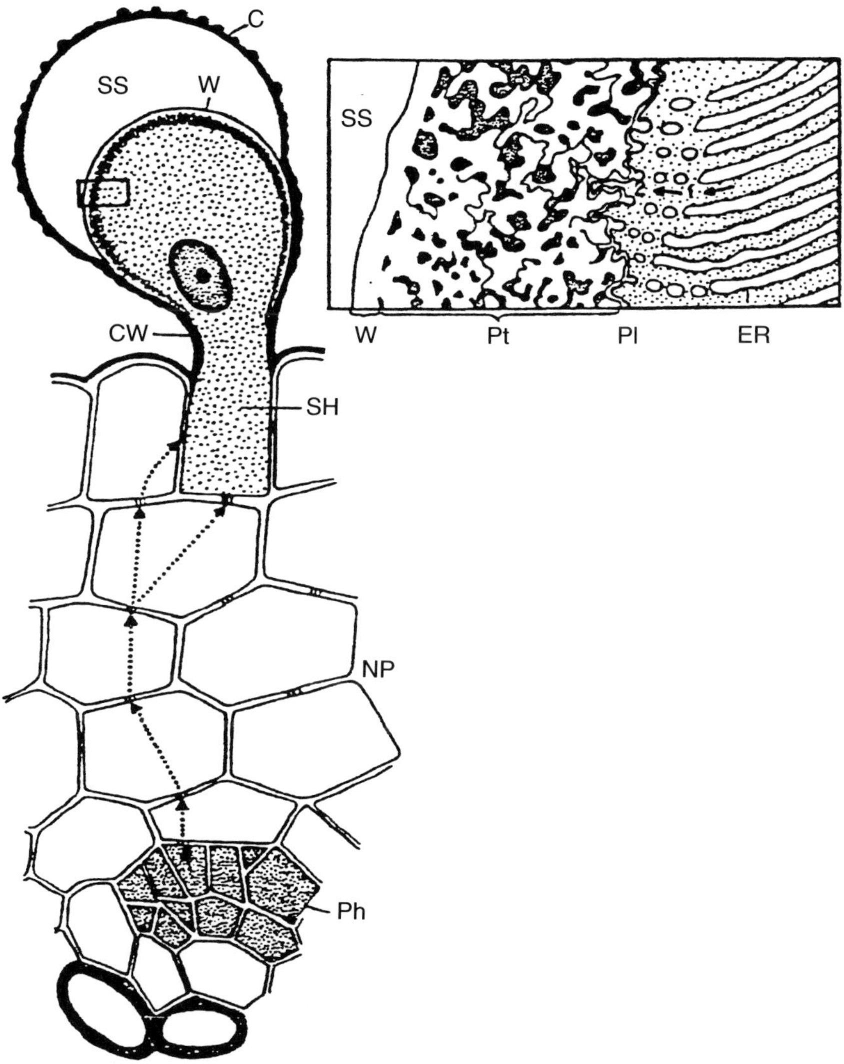

Fig. 4. A portion of the nectary of *Lonicera japonica* illustrating a mode of secretion as suggested by Fahn (1979*a*). The large figure shows the flow of prenectar from the phloem (Ph) via plasmodesmata to the secretory hair (SH). The inset provides a possibility for the means of nectar secretion from the protoplast. It is envisaged that vesicles are secreted from the ER and fuse with the plasmalemma (Pl) from where the nectar moves through wall protuberances (Pt) and primary wall (W) into the subcuticular space (SS). CW, cutinised wall; C, cuticle; NP, nectar pathway (plasmodesmata); ER, endoplasmic reticulum. (Redrawn from Fahn 1979*b*.)

formed when nectar secretion was inhibited; and the efflux rate (0.3 μM saccharose s^{-1} (m plasmalemma)$^{-2}$) was high but still compatible with an eccrine process. In the case of exocytosis, however, it was calculated that much more membrane material was incorporated into the plasmalemma than would be necessary for the formation of the protuberances alone.

Nectar is exuded via stomata from the floral nectaries of many plants. *Tropaeolum* has such nectaries, where the exudate is released into intercellular spaces of a secretory parenchyma, flows to the lowest portion of the spur and is exuded via the modified stomata. Granulocrine secretion, with the vesicles derived from both Golgi bodies and ER cisternae, has been proposed (see Table 1 for references). It has also been suggested that the secretory vesicles in *Citrus sinensis* and *Vinca* (*rosea* and *major*) are derived from the ER (see Table 1).

Extrafloral nectaries

Extrafloral nectaries occur on vegetative (and sometimes reproductive) parts of plants. They are generally interpreted to be important in maintaining a mutually beneficial relationship between the particular plant and certain insects, which are attracted to the nectar and in turn offer the plant varying degrees of protection. The nectar contains sugars (mainly glucose, fructose and sucrose) and organic compounds (e.g. amino acids).

The sugar concentration of the nectar from the extrafloral nectaries of *Vicia faba* was low in comparison with that of floral nectaries in the same species (Davis *et al.* 1988). Secretion was assumed to be granulocrine with the vesicles derived from the ER in this plant as well as in *Plumeria*, *Gossypium* and *Impatiens* (Table 1).

Eccrine secretion was postulated for the floral and extrafloral nectaries of *Passiflora* spp. by Durkee (1981, 1982, 1983) as well as for the extrafloral nectaries of *Ancistrocactus scheeri* (Mauseth 1982). *Vigna unguiculata* (Kuo & Pate 1985) and *Ricinus communis* (Baker *et al.* 1978). The nectaries of *Acacia* differ between species and both granulocrine and eccrine mechanisms have been proposed (Marginson *et al.* 1985*a*), although in *A. terminalis* Marginson *et al.* (1985*b*) took the view that only granulocrine secretion occurred.

Stinging trichomes

Stinging trichomes are deterrents as they cause pain, when touched, due to the release of toxic substances. Marty (1968) studied the ultrastructure of secretory trichomes of *Urtica urens* and mentions two possibilities: (1) that the Golgi bodies transfer substances from the ER to the vacuoles;

and (2) that there may be continuity between the ER cavities and the vacuoles, with the direct transfer of the secreted products as a result.

Lipophilic secretion

Lipophilic materials secreted by plant glands comprise a variety of different substances including terpenes, waxes and fats. Many of them are produced in glands of the holocrine type or start in a merocrine manner and terminate in a holocrine way. The ER is typically smooth and tubular in cells of this kind.

Kroh (1967), Fahn & Benayoun (1976) and Benayoun & Fahn (1979) are among those who have proposed that lipophilic substances leave the protoplast either by reverse pinocytosis or by droplets becoming completely surrounded by plasmalemma invaginations (Fig. 5).

Glandular trichomes

Oil secretion has been interpreted to be either granulocrine (Table 1), or eccrine as in the bud trichomes of *Alnus* (Wollenweber *et al.* 1971) and in the farina secretions from glands of primroses (Wollenweber & Schnepf 1970).

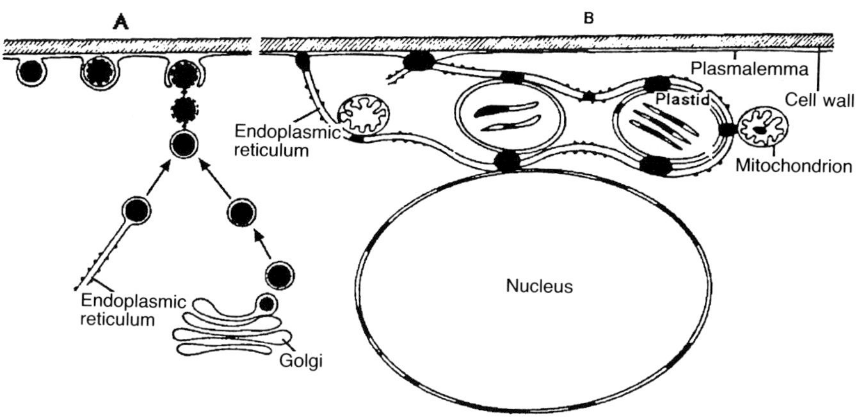

Fig. 5. Possible mechanisms for the elimination of resin from the protoplasts of epithelial duct cells of *Pinus halpensis*. (a) Vesicles of Golgi or ER origin containing resin are eliminated by plasmalemma invaginations. (b) ER is shown to collect resin droplets from various organelles, transport them towards the plasmalemma and eliminate them at the point of contact. Fusion of ER and plasmalemma are assumed to occur. (Redrawn from Benayoun & Fahn 1979.)

Glandular areas in the reproductive system

The wet stigma is a gland evolved to facilitate pollination and consists of specialised receptive cells which are able to recognise and discriminate between pollen grains according to their genotype (Dumas *et al.* 1988). The secretion is often complex and can contain sugars, lipids, proteins, and other components. The nature of the exudates appears to vary widely between different species, with two main secretory mechanisms: granulocrine secretion for hydrophilic substances, and holocrine secretion in stigmas with a lipophilic exudate (Dumas *et al.* 1988).

An ER-derived granulocrine secretion is assumed in some plants (see Table 1 for references). Dumas *et al.* (1978) proposed simultaneous eccrine and exocytotic pathways in the stigma of *Lycopersicum peruvianum*, a concept that they felt was supported by the observed heterogeneous nature of the exudate. Heslop-Harrison & Heslop-Harrison (1980) proposed the involvement of paramural bodies or multivesicular bodies in the granulocrine secretion of proteins in the stigmatic papillae of *Secale* and *Hordeum*, while lipidic secretion was assumed to be eccrine in the stigma of *Oenothera organensis* (Heslop-Harrison & Heslop-Harrison 1985) and in avocado (Sedgley & Blesing 1983), lipid or lipid precursors being synthesised and transferred by mechanisms that were not structurally visible.

The placenta of hot and sweet cultivars of red pepper (*Capsicum annuum*) has glandular areas. ER fragments were interpreted to fuse with the plasmalemma (Zamski *et al.* 1987) while Kristen (1976) assumed an exocytotic extrusion mediated by Golgi vesicles in the placental papillae of *Aptenia cordifolia* (Fig. 6). He also suggested the possibility of holocrine secretion, in addition to the exocytosis.

Osmophores

The osmophore is a glandular floral tissue that attracts pollinators by emitting a fragrance comprising terpenoids and aromatics. Odorous oils penetrate through the plasmalemma as molecules or micelles (Curry *et al.* 1988/9).

Both granulocrine (Table 1) and eccrine secretion have been reported. During the period of aroma emission, the secretion of lipid droplets, according to Schnepf (1965*c*), was eccrine in *Typhonium divaricatum* (Arecaceae), while different species of the orchid *Stanhopea* (*anfracta* and *pulla*) were considered to exhibit granulocrine and eccrine secretion, respectively. These two species are thought to be at opposite poles in the evolutionary development of the genus (Stern *et al.* 1987; Curry *et al.* 1988/9).

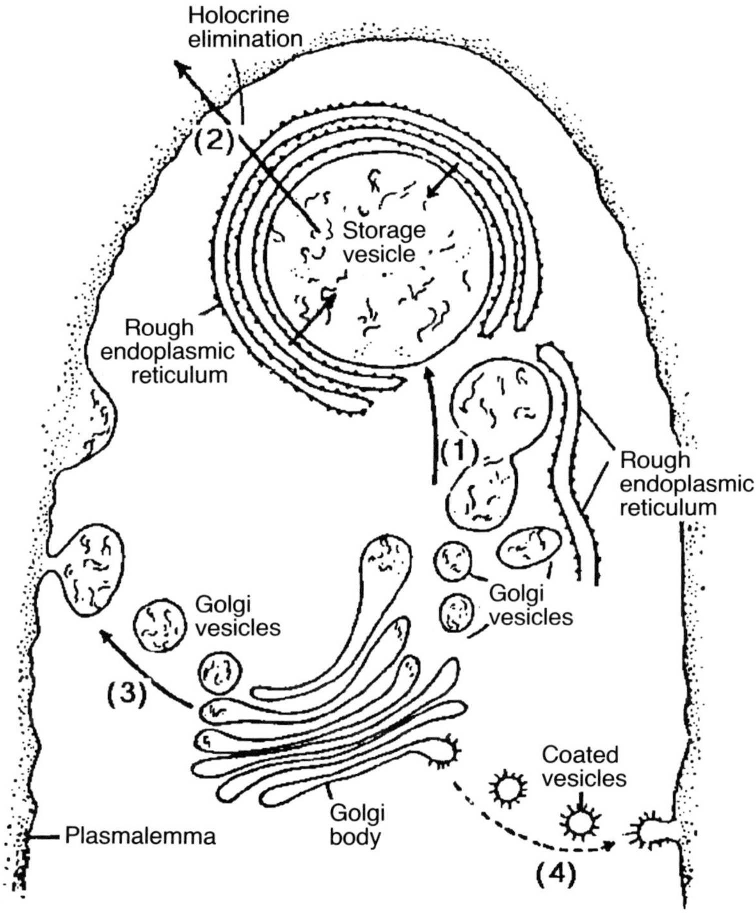

Fig. 6. Model of the manner of secretion of protein-carbohydrate mucilage by placental papillae of *Aptenia cordifolia* suggested by Kristen (1976). 1, Golgi vesicles fuse to form a storage vesicle which becomes surrounded by rough endoplasmic reticulum (RER) – from the RER proteins enter the storage vesicle; 2, holocrine elimination by degeneration of protoplast; 3, Golgi vesicles containing polysaccharide mucilage eliminate it by fusion with the plasmalemma; 4, possible secretion of polysaccharides by coated vesicles. (Redrawn from Fahn 1979*b*.)

Secretory processes including some models

The purpose of the above brief review has been to demonstrate that, although a wide variety of glands has been studied, we still have remarkably little information about the actual transport processes that are taking place. All too often, conclusions as to the route and mechanisms of secretion have relied on shallow observational evidence coupled with subjective interpretation.

Eccrine secretion

It is not easy to *prove* the operation of an eccrine process of secretion which is, therefore, more often deduced by means of exclusion of granulocrine mechanisms. Most published results have relied on the absence of ER-derived or Golgi vesicles at the relevant time, sometimes together with the histochemical demonstration of active metabolic processes at the plasmalemma. Firm evidence can, however, be obtained only by following the pathway of specific molecules or ions. There remains much to do in this area and, given the substantial work load in studying even a single gland adequately, it is likely to be some considerable time before we have sufficient evidence to draw any useful, unifying hypotheses on secretory mechanisms.

Exocytotic (granulocrine) secretion

In general, it has been easier to find support evidence for granulocrine secretion than for eccrine transport. In particular, some ultrastructural manifestation of exocytosis is almost always obvious. Nevertheless, and as already commented, structural studies of nectaries and other glands have too often attempted to deduce functional mechanisms from purely ultrastructural observations. Only rarely have physiological data been incorporated with quantitative structural information so that the relative merits of alternative mechanisms can be adequately explored. In some instances even a superficial quantitative assessment would have prevented authors from making suggestions that could easily be proved to be impossible on physiological grounds. Such reasoning led Robards & Stark (1988) to review previous work on the nectary of *Abutilon* (Findlay & Mercer 1971*a,b*; Findlay *et al.* 1971; Reed *et al.* 1971; Gunning & Hughes 1976; Hughes 1977; Robards 1984, 1985; Robards & Oates 1986; Kronestedt *et al.* 1986; Terry & Robards 1987) and to integrate it with new ultrastructural and physiological evidence. Even in this much-studied nectary, the actual mechanism of secretion remained unresolved after almost 20 years of quite detailed study. The 'per trichome' volume secretion rate has previously been determined by a number of authors and is usually within the range 1.0 to 100 $\mu m^3\ s^{-1}$. Robards & Stark (1988) used a mid-range figure of 50 $\mu m^3\ s^{-1}$ in their calculations (= approximately 1.0×10^{-11} g sugar s^{-1}). To secrete this amount of nectar would require a transmembrane flux of about 85×10^{-6} mol s^{-1} m^{-2} if the whole tip cell membrane only were involved or 3×10^{-6} to 4×10^{-6} mol s^{-1} m^{-2} if the total plasmalemma surface area of the hair were available for unloading. These rates of eccrine secretion appear high compared with other transmembrane sugar fluxes found in plants, although less so when compared with some ion fluxes (see Kronestedt & Robards 1987). With regard to granulocrine secretion, calculation of the required rate of vesicle fusion to sustain the necessary volume secretion rate revealed that, for 50 nm diameter vesicles, no less than 750 000 would need to fuse with the tip cell membrane every second while, if the vesicles were 500 nm in diameter, the fusion rate would still be about 750 s^{-1}. In either case, cell membrane extension and the consequential need for membrane recycling would appear to impose impossible demands on the cell. Even if vesicle fusion were to occur in all the cells of the hair, the relevant rates would be reduced by a factor of only about 20. These calculations, derived from existing physiological data, did not provide an unambiguous explanation of the processes of secretion in *Abutilon* and Robards & Stark (1988) therefore sought additional information from electron micrographs of rapidly frozen secreting hairs.

Rapid freezing followed by freeze-substitution or freeze-fracture replication revealed an extensive 'secretory reticulum' (SR) present within the hair cells (Fig. 7). As similar micrographs were obtained from conventional, chemical, fixation it was argued that the SR is a relatively stable endomembrane system. Freeze-fracture and freeze-substitution micrographs showed that this internal membrane system is closely associated with the plasmalemma. The actual surface area of SR within a single trichome is at least 50 000 μm^2, which would mean that the total surface area of SR in the hairs of a mature gland would probably be more than 100 mm × 100 mm! Taken together with other structural information, as well as physiological data, it was concluded that prenectar is actively loaded across the ample membrane area into the SR of all trichome cells. The transmembrane flux required would only be 6.0×10^{-7} mol m^{-2} s^{-1}. It was postulated that an increase in the hydrostatic pressure within this compartment leads to the opening of 'sphincters' which connect the cisternal space of the SR to the outside of the plasmalemma. Thus a pulse of nectar would be forcibly expelled into an apoplastic compartment sealed to the outside by the impermeable cuticle and on the inside by the plasmalemma. As this apoplastic compartment is also sealed at the stalk cell, the only route for pressure release is via the transient pores which overlay the tip cell. Distension renders these patent so that, again, pulsed secretion is observed. This hypothesis overcomes the necessity for envisaging excessively high transmembrane fluxes or rates of vesicle fusion. It

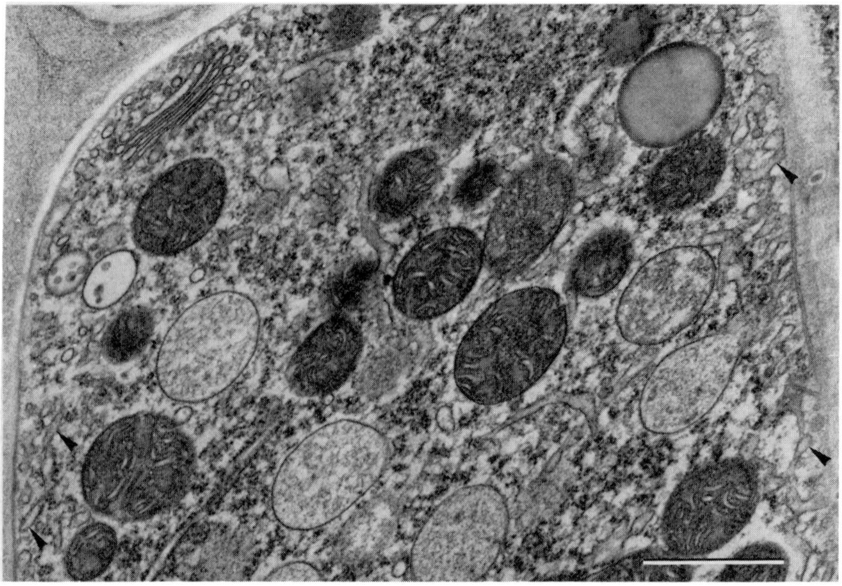

Fig. 7. Freeze-substituted hair cell of a secreting *Abutilon* gland. There is little, if any, evidence of disruption from ice crystals formed during freezing or during the substitution process. The section has cut through the cell walls rather obliquely, so revealing underlying microtubules and anastomosing cisternae of secretory reticulum (arrowed). Deeper within the cell, dictyosome stacks are seen. The bar represents 1.0 μm.

would also imply the need for a continuing supply of prenectar to the hair cells accompanied by active loading into the SR. This loading process may well be supported by the hydrolysis of sucrose to glucose and fructose and is probably the site where ions and other low molecular weight solutes are filtered from the nectar. This model is illustrated in Fig. 8. It will be seen that it is certainly not an eccrine process but neither is it a classical granulocrine secretion. However, it does accord with all the evidence currently available and overcomes the quantitative limitations apparently imposed by invoking normal eccrine or granulocrine processes alone. From this brief case-study we can learn that it is only when we have adequate structural *and* physiological data that we can combine the two and make sensible judgements on actual mechanisms. By contrast, the ultrastructural appearance of the *Fritillaria* nectary (Fig. 9) indicates a much clearer, and apparently permanent, association of the plasmalemma with vesicular extrusion.

(a)

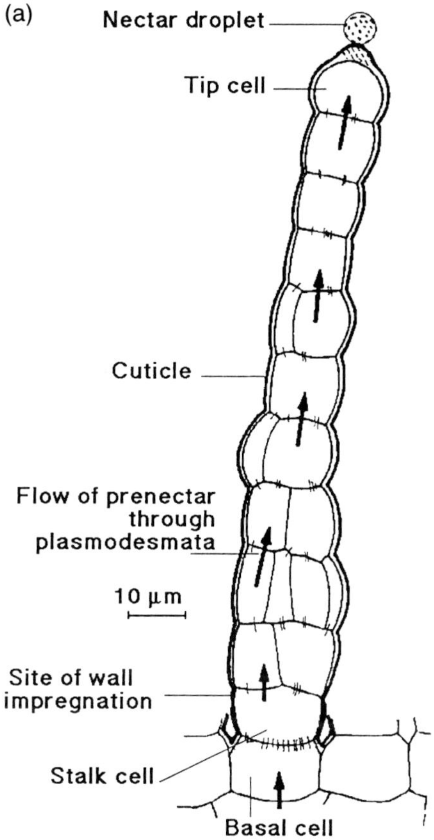

Fig. 8 (a)–(c). Diagrams illustrating the model for secretion based on the evidence from freeze-substitution, freeze-fracture and physiological experiments. It is envisaged that prenectar moves into the symplast of the hair via the numerous plasmodesmata in the transverse walls of the stalk cell. In each of the hair cells some of the prenectar is loaded from the cytoplasm into the secretory reticulum. It is at this stage that a filtration effect takes place, so defining the chemical composition of the secreted product. The sucrose is partially hydrolysed to glucose and fructose but evidence is not yet available as to whether this takes place at the membrane or within the cisternal cavity of the secretory reticulum. As loading into the secretory reticulum continues, a hydrostatic pressure builds up until, ultimately, a minute pulse of nectar is forced into the freely permeable apoplastic space bounded by the plasmalemma to the interior and the cuticle to the exterior. The continuing build-up of pressure within this compartment ultimately reaches the level where pores in the cuticle over the tip cell become patent and release a pulse of nectar

(b)

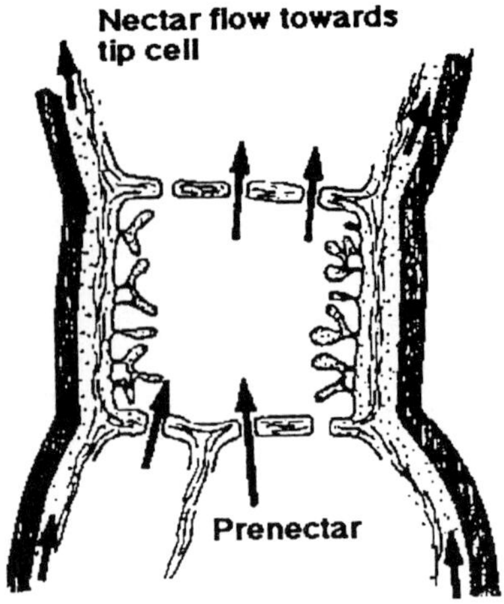

(c)

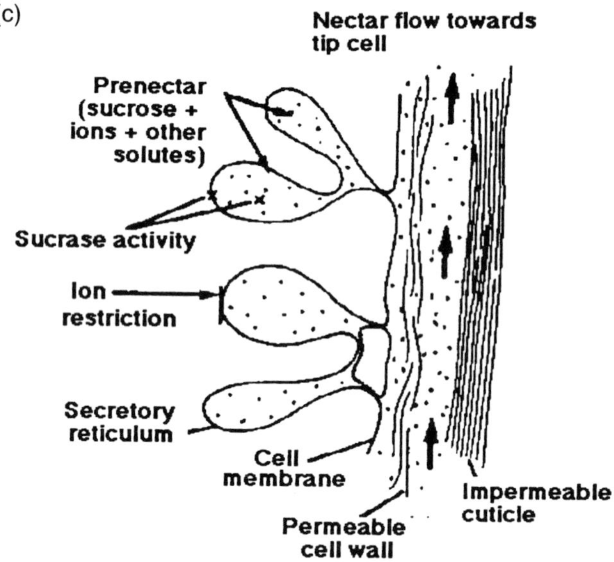

Fig. 8 (*contd*)

to the exterior. Such a model takes account of all the currently available information concerning secretion in *Abutilon* but does not predict excessively high rates of either eccrine or (conventional) granulocrine secretion. It also overcomes the osmotic difficulties that would otherwise arise from the high concentrations of solutes that are transported in this system. (From Robards & Stark 1988.)

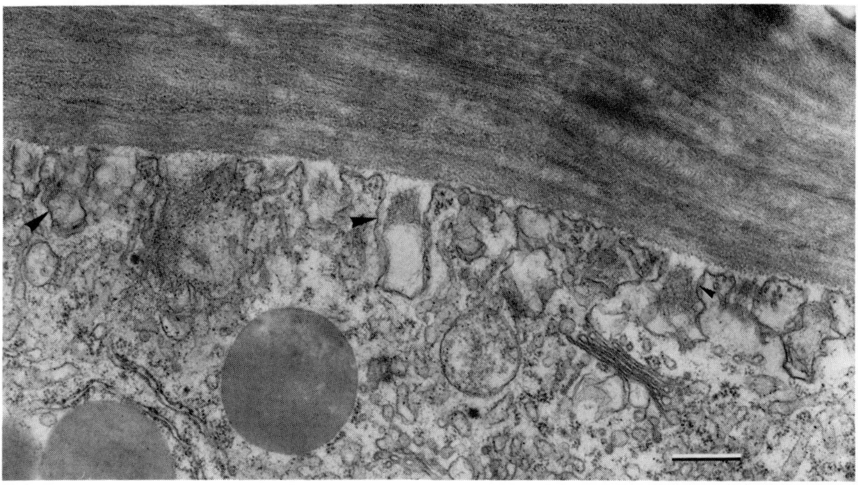

Fig. 9. Nectary of *Fritillaria*. Arrowheads indicate where exocytotic extrusion apparently takes place through the plasmalemma towards the exterior. The bar represents 0.5 μm.

Conclusions

At least some representatives have been investigated from most exogenous types of gland. Interpretation of *how* the secretion is released still appears to be very subjective. If vesicles have been found close to the plasmalemma, interpretation has favoured granulocrine secretion and the source of vesicles (ER or Golgi) has then often been postulated. Apart from identifying cellular compartments, a further complication involves the dynamics of the cell: it is impossible to state from micrographs in which direction the vesicles are moving. Perhaps a presumed exocytosis is actually an endocytotic event in progress! Another potential cause of misinterpretation is the possibility that fusion of different membranes may have taken place during the course of specimen preparation. If, on the other hand, no vesicles have been found, this has been taken as a support for eccrine secretion, especially if wall protuberances are present in the cell in the area of secretion elimination.

It appears to be necessary to define which features should be looked for when analysing secretory events. Cytological investigations must naturally be combined with physiological analytical methods, as Schnepf with collaborators and Robards in particular have done. If vesicles are present, their contents should be identified and this should be related to type of secretion and activity of the plasmalemma together with the secretion rate.

The most general conclusion that can be drawn is that both eccrine and granulocrine secretion are possible (cf. Lüttge & Schnepf 1976) and that there may be quite elaborate modifications (e.g. in *Abutilon*) of what have previously been considered to be relatively simple pathways.

Summary

Plant glands comprise groups of highly specialised cells capable of discharging one or more of a wide range of secretory products to the outside of the protoplast. Glands are usually positioned superficially and vary in morphology from simple cellular protuberances to extensive aggregations of quite elaborate hairs (trichomes). Secretory processes have classically been divided into two major categories: eccrine and granulocrine. Eccrine secretion involves the direct transport of ions or molecules across the cell membrane, while in granulocrine secretion cytoplasmic vesicles fuse with the cell membrane, so releasing their contents to the exterior. Granulocrine secretion thus incorporates the process of exocytosis, the specific subject of this chapter.

A wide range of glands has been described in the literature, the major secretory products including mucilages, salts, sugars and oils. In only a few cases has a gland been studied in sufficient detail to allow a full statement to be made about the pathway and mechanism of secretion. However, similar substances may be discharged by either eccrine or granulocrine secretion from different plants. Although our knowledge is incomplete, it is clear that this relates both to the evolutionary status of the plant as well as to the nature of the secretory product and, crucially, to the actual secretion *rate*.

Nectaries are among the best-studied glands: eccrine secretion has been demonstrated and may be aided by extensive membrane amplification. Clearly there is an upper limit to the rate at which transmembrane transport may take place; similarly the rate of fusion, and recycling, of exocytotic vesicles places limits on the capacity of granulocrine secretion. Measurements of secretion rates in some glands, such as the *Abutilon* nectary, suggest that secretion must be by a modified exocytotic mechanism.

References

Allison, A.C. & Davies, P. (1974). Mechanisms of endocytosis and exocytosis. *Symposia of the Society for Experimental Biology*, **28**, 419–46.

Amelunxen, F. (1964). Elektronenmikroskopische Untersuchungen an den Drüsenhaaren von *Mentha piperita* L. *Planta Medica*, **12**, 121–39.

Baker, D.A., Hall, J.L. & Thorpe, J.R. (1978). A study of the extrafloral nectaries of *Ricinus communis*. *New Phytologist*, **81**, 129–37.

Belin-Depoux, M. & Clair-Maczulajtys, D. (1975). Introduction à l'étude des glandes foliaires de l'*Aleureites moluccana* Willd. II. Aspects histologiques de la glande pétiolaire fonctionnelle. *Revue générale de Botanique*, **82**, 119–55.

Benayoun, J. & Fahn, A. (1979). Intracellular transport and elimination of resin from epithelial duct-cells of *Pinus halepensis*. *Annals of Botany*, **43**, 179–81.

Benner, U. & Schnepf, E. (1975). Die Morphologie der Nektarausscheidung bei Bromeliaceen: Beteiligung des Golgi-Apparates. *Protoplasma*, **85**, 337–49.

Christ, P. & Schnepf, E. (1988). Zur Struktur und Funktion von Asclepiadaceen-Nektarien. *Beiträge zur Biologie der Pflanzen*, **63**, 55–80.

Ciobanu, I.R. & Tacina, F. (1973). L'ultrastructure des glandes digestives chez *Drosera capensis* L. *Revue roumaine de Biologie*, **18**, 249–53.

Curry, K.J., Stern, W.L. & McDowell, L.M. (1988/9). Osmophore development in *Stanhopea anfracta* and *S. pulla* (Orchidaceae). *Lindleyana*, **3**, 212–20.

Davis, A.R., Peterson, R.L. & Shuel, R.W. (1986). Anatomy and vasculature of the floral nectaries of *Brassica napus* (Brassicaceae). *Canadian Journal of Botany*, **64**, 2508–16.

Davis, A.R., Peterson, R.L. & Shuel, R.W. (1988). Vasculature and ultrastructure of the floral and stipular nectaries of *Vicia faba* (Leguminosae). *Canadian Journal of Botany*, **66**, 1435–48.

Dexheimer, J. (1976). Etude de la secretion de mucilage par les cellules des glandes digestives de *Drosera* (*D. rotundifolia* L.; et *D. capensis* L.). Application de quelques techniques cytochimiques. *Cytobiologie*, **13**, 307–21.

Dumas, C. (1973). Contribution à l'étude cyto-physiologique du stigmate: III. Evolution et rôle du réticulum endoplasmique au cours de la sécrétion chez *Forsythia intermedia*; étude cytochimique. *Zeitschrift für Pflanzenphysiologie*, **70**, 119–30.

Dumas, C., Bowman, R.B., Gaude, T., Guilly, C.M., Heizmann, P., Roeckel, P. & Rougier, M. (1988). Stigma and stigmatic secretion reexamined. *Phyton (Horn)*, **28**, 193–200.

Dumas, C., Rougier, M., Zandonella, P., Ciampolini, F., Cresti, M. & Pacini, E. (1978). The secretory stigma in *Lycopersicum peruvianum* Mill.: Ontogenesis and glandular activity. *Protoplasma*, **96**, 173–87.

Durkee, L.T. (1981). The floral and extrafloral nectaries of *Passiflora*. I. The floral nectary. *American Journal of Botany*, **68**, 453–62.

Durkee, L.T. (1982). The floral and extrafloral nectaries of *Passiflora*. II. The extrafloral nectary. *American Journal of Botany*, **69**, 1420–8.

Durkee, L.T. (1983). The ultrastructure of floral and extrafloral nectaries. In *The Biology of Nectaries*, ed. B. Bentley & T. Elias, pp. 1–29. Columbia: The University Press.

Eleftheriou, E.P. & Hall, J.L. (1983*a*). The extra-floral nectaries of cotton. I. Fine structure of the secretory papillae. *Journal of Experimental Botany*, **34**, 103–19.

Eleftheriou, E.P. & Hall, J.L. (1983*b*). The extra-floral nectaries of cotton. II. Cytochemical localization of ATPase activity and Ca^{2+}-binding sites, and selective osmium impregnation. *Journal of Experimental Botany*, **34**, 1066–79.

Eriksson, M. (1977). The ultrastructure of the nectary of red clover (*Trifolium pratense*). *Journal of Apicultural Research* (*Eng.*), **16**, 184–93.

Eymé, J. (1966). Infrastructure des cellules nectarigènes des *Diplotaxis erucoides* D.C., *Helleborus niger* L., et *Helleborus foetidus* L. *Comptes Rendus hebdomadaires des Séances de l'Académie des Sciences. Serie D*, **262**, 1629–32.

Eymé, J. & LeBlanc, M. (1963). Contribution à l'étude inframicroscopique d'inclusions cytoplasmiques présentes dans les ovules de *Ficaria* et dans les nectaires d'*Helleborus*. *Comptes Rendus hebdomadaires des Séances de l'Académie des Sciences. Serie D*, **256**, 4958–9.

Fahn, A. (1949). Studies in the ecology of nectar secretion. *Palestine Journal of Botany, Jerusalem Series*, **4**, 207–24.

Fahn, A. (1952). On the structure of floral nectaries. *Botanical Gazette*, **113**, 464–70.

Fahn, A. (1979*a*). Ultrastructure of nectaries in relation to nectar secretion. *American Journal of Botany*, **66**, 977–85.

Fahn, A. (1979*b*). *Secretory Tissues in Plants*. London: Academic Press.

Fahn, A. (1987). Secretory tissues in plants. *Israel Journal of Botany*, **36**, 199–213.

Fahn, A. (1988*a*). Tansley review no. 14. Secretory tissues in vascular plants. *New Phytologist*, **108**, 229–57.

Fahn, A. (1988*b*). Secretory tissues and factors influencing their development. *Phyton* (*Horn*), **28**, 13–26.

Fahn, A. & Benayoun, J. (1976). Ultrastructure of resin ducts in *Pinus halepensis*. Development, possible sites of resin synthesis, and mode of its elimination from the protoplast. *Annals of Botany*, **40**, 857–63.

Fahn, A. & Benouaiche, P. (1979). Ultrastructure, development, and secretion in the nectary of banana flowers. *Annals of Botany*, **44**, 85–93.

Fahn, A. & Rachmilewitz, T. (1970). Ultrastructure and nectar secretion in *Lonicera japonica*. *Journal of the Linnean Society, London*, **63**, 51–6.

Fahn, A. & Rachmilevitz, T. (1975). An autoradiographical study of nectar secretion in *Lonicera japonica* Thunb. *Annals of Botany*, **39**, 975–6.

Faraday, C. D. & Thomson, W. W. (1986). Morphometric analysis of *Limonium* salt glands in relation to ion efflux. *Journal of Experimental Botany*, **37**, 471–81.

Figier, J. (1972). Localisation infrastructurale de la phosphatase acide dans les glandes pétiolaires d'*Impatiens holstii*. *Planta*, **108**, 215–26.

Findlay, N. & Mercer, F. V. (1971*a*). Nectar production in *Abutilon*. 1. Movement of nectar through the cuticle. *Australian Journal of Biological Sciences*, **24**, 647–56.

Findlay, N. & Mercer, F. V. (1971*b*). Nectar production in *Abutilon*. 2. Sub-microscopic structure of the nectary. *Australian Journal of Biological Sciences*, **24**, 657–64.

Findlay, N., Reed, M. & Mercer, F. V. (1971). Nectar production in *Abutilon*. 3. Sugar secretion. *Australian Journal of Biological Sciences*, **24**, 665–75.

Fiordi, A. C. & Palandri, M. R. (1982). Anatomic and ultrastructural study of the septal nectary in some *Tillandsia* (Bromeliaceae) species. *Caryologia*, **35**, 477–89.

Gilchrist, A. J. & Juniper, B. E. (1974). An excitable membrane in the stalked glands of *Drosera capensis* L. *Planta*, **119**, 143–7.

Gunning, B. E. S. & Hughes, J. E. (1976). Quantitative assessment of symplastic transport of pre-nectar into trichomes of *Abutilon* nectaries. *Australian Journal of Biological Sciences*, **3**, 619–37.

Healey, P. L. (1967). The secretory trichomes of *Pharbitis* – its ultrastructural and histochemical development. *American Journal of Botany*, **54**, 638.

Heinrich, G. (1975*a*). Über den Glucose-Metabolismus in Nektarien zweier *Aloe*-Arten und über den Mechanismus der Pronektar-Sekretion. *Protoplasma*, **85**, 351–71.

Heinrich, G. (1975*b*). Über die Lokalisation verschiedener Phosphatasen im Nektarium von *Aloe*. *Cytobiologie*, **11**, 247–63.

Heslop-Harrison, J. & Heslop-Harrison, Y. (1980). The pollen–stigma interaction in the grasses. I. Fine structure and cytochemistry of the stigmas of *Hordeum* and *Secale*. *Acta Botanica Neerlandica*, **29**, 261–76.

Heslop-Harrison, J. & Heslop-Harrison, Y. (1985). The secretory system of the stigma of *Oenothera organensis* Munz: some developmental and quantitative features. *Israel Journal of Botany*, **34**, 187–204.

Hill, A. E. & Hill, B. S. (1973). The *Limonium* salt gland: a biophysical and structural study. *International Review of Cytology*, **35**, 229–320.

Hill, A. E. & Hill, B. S. (1976). Elimination processes by glands. Mineral ions. In *Encyclopedia of Plant Physiology*, new series, vol. 2, *Transport in Plants*, II B, ed. U. Lüttge & M. G. Pitman, pp. 225–43. Berlin: Springer-Verlag.

Horner, H. T. & Lersten, N. R. (1967). Development, structure and function of dendroid secretory trichomes in *Psychotria bacteriophila* (Rubiaceae). *American Journal of Botany*, **56**, 638.

Horner, H.T. & Lersten, N.R. (1968). Development, structure and function of secretory trichomes in *Psychotria bacteriophila* (Rubiaceae). *American Journal of Botany*, **55**, 1089–99.

Hughes, J.E. (1977). Aspects of ultrastructure and function in *Abutilon* nectaries. M.Sc. thesis, Australian National University.

Kristen, U. (1976). Die Morphologie der Schleimsekretion in Fruchtknoten von *Aptenia cordifola*. *Protoplasma*, **89**, 221–33.

Kristen, U. (1977). Granulocrine Ausscheidung von Narbensekret durch Vesikel des Endoplasmatischen Retikulums bei *Aptenia cordifolia*. *Protoplasma*, **92**, 243–51.

Kristen, U. (1978). Granulokrine Sekretion bei Pflanzen. *Biologische Rundschau*, **16**, 365–75.

Kristen, U., Liebezeit, G. & Biedermann, M. (1982). The ligule of *Isoetes lacustris*: ultrastructure, mucilage composition and a possible pathway of secretion. *Annals of Botany*, **49**, 569–84.

Kristen, U. & Lockhausen, J. (1985). The leaf glands of *Veronica beccabunga* L.: ultrastructure and a possible pathway of secretion. *Israel Journal of Botany*, **6**, 147–56.

Kroh, M. (1967). Bildung und Transport des Narbensekrets von *Petunia hybrida*. *Planta*, **77**, 250–60.

Kronestedt, E.C. & Robards, A.W. (1987). Sugar secretion from the nectary of *Strelitzia*: an ultrastructural and physiological study. *Protoplasma*, **137**, 168–82.

Kronestedt, E.C., Robards, A.W., Stark, M. & Olesen, P. (1986). Development of trichomes in the *Abutilon* nectary gland. *Nordic Journal of Botany*, **6**, 627–39.

Kronestedt, E. & Walles, B. (1986). Anatomy of the *Strelitzia reginae* flower. *Nordic Journal of Botany*, **6**, 307–20.

Kronestedt-Robards, E.C., Greger, M. & Robards, A.W. (1989). The nectar of the *Strelitzia reginae* flower. *Physiologia Plantarum*, **77**, 341–6.

Kuo, J. & Pate, J.S. (1985). The extrafloral nectaries of cowpea (*Vigna unguiculata* (L.) Walp) I. Morphology, anatomy, and fine structure. *Planta*, **166**, 15–27.

Lüttge, U. (1961). Uber die Zusammensetzung des Nektars und den Mechanismus seiner Sekretion. I. *Planta*, **56**, 189–212.

Lüttge, U. (1971). Structure and function of plant glands. *Annual Review of Plant Physiology*, **22**, 23–44.

Lüttge, U. & Schnepf, E. (1976). Organic substances. In *Encyclopedia of Plant Physiology*. New series, vol. 2, *Transport in Plants*, part B, ed. U. Lüttge & M.C. Pitman, pp. 244–77. New York: Springer-Verlag.

Marginson, R., Sedgley, M., Douglas, T.J. & Knox, R.B. (1985*a*). Structure and secretion of the extrafloral nectaries of Australian acacias. *Israel Journal of Botany*, **34**, 91–102.

Marginson, R., Sedgley, M. & Knox, R.B. (1985*b*). Structure and histochemistry of the extrafloral nectary of *Acacia terminalis* (Salisb.) MacBride (Leguminosae, Mimosoideae). *Protoplasma*, **127**, 21–30.

Marty, M.F. (1968). Infrastructures des organes sécréteurs de la feuille d'*Urtica urens*. *Comptes Rendus hebdomadaires des Séances de l'Académie des Sciences, Serie D*, **266**, 1712–14.

Mauseth, J.D. (1982). Development and ultrastructure of extrafloral nectaries in *Ancistrocactus scheeri* (Cactaceae). *Botanical Gazette*, **143**, 273–7.

Meyberg, M. & Kristen, U. (1981). The nectaries *Aptenia cordifolia*. Ultrastructure, translocation of ^{14}C-labelled sugars, and a possible pathway of secretion. *Zeitschrift für Pflanzenphysiologie*, **104**, 139–47.

Mohan, J.S.S. & Inamdar, J.A. (1986). Ultrastructure and secretion of extrafloral nectaries of *Plumeria rubra* L. *Annals of Botany*, **57**, 389–401.

Mollenhauer, H.H. (1967). The fine structure of mucilage secreting cells of *Hibiscus esculentus* pods. *Protoplasma*, **63**, 353–62.

Osmond, C.B., Lüttge, U., West, K.R. & Pallaghy, C.K. (1969). Ion absorption in *Atriplex* leaf tissue. II. Secretion of ions to epidermal bladders. *Australian Journal of Biological Sciences*, **22**, 797–814.

Perrin, A. (1970). Organisation infrastructurale, en rapport avec les processus de sécrétion, des poils glandulaires (trichome-hydathodes) de *Phaseolus multiflorus*. *Comptes Rendus hebdomadaires des Séances de l'Académie des Sciences, Serie D*, **270**, 1984–7.

Perrin, A. (1971). Présence de 'cellules de transfer' au sein de l'épithème de quelques hydathodes. *Zeitschrift für Pflanzenphysiologie*, **65**, 39–51.

Pridgeon, A.M. & Stern, W.L. (1983). Ultrastructure of osmophores in *Restrepia* (Orchidaceae). *American Journal of Botany*, **70**, 1233–43.

Rachmilevitz, T. & Fahn, A. (1973). Ultrastructure of nectaries of *Vinca rosea* L., *Vinca major* L. and *Citrus sinensis* Osbeck cv. Valencia and its relation to the mechanism of nectar secretion. *Annals of Botany*, **37**, 1–9.

Rachmilevitz, T. & Fahn, A. (1975). The floral nectary of *Tropaeolum majus* L. The nature of the secretory cells and the manner of nectar secretion. *Annals of Botany*, **39**, 721–8.

Reed, M.L., Findlay, N., Mercer, F.V. (1971). Nectar production in *Abutilon*. 4 – Water and solute relations. *Australian Journal of Biological Sciences*, **24**, 677–88.

Robards, A.W. (1984). Fact or artefact – a cool look at biological electron microscopy. (Presidential address to the Microscopical Society). *Proceedings of the Royal Microscopical Society*, **19**, 195–208.

Robards, A.W. (1985). The use of low temperature methods for structural and analytical studies of plant transport processes. In *Botanical Microscopy 1985*, ed. A.W. Robards, pp. 39–64. Oxford: Oxford University Press.

Robards, A.W. & Oates, K. (1986). X-ray microanalysis of ion distribution in *Abutilon* nectary hairs. *Journal of Experimental Botany*, **37**, 940–6.

Robards, A.W. & Stark, M. (1988). Nectar secretion in *Abutilon*: a new model. *Protoplasma*, **142**, 79–91.

Robins, R.J. & Juniper, B.E. (1980*a*). The secretory cycle of *Dionaea muscipula* Ellis. II. Storage and synthesis of the secretory proteins. *New Phytologist*, **86**, 297–311.

Robins, R.J. & Juniper, B.E. (1980*b*). The secretory cycle of *Dionaea muscipula* Ellis. III. The mechanism of release of digestive secretion. *New Phytologist*, **86**, 313–27.

Rougier, M. (1969). Mise en évidence de la fonction sécrétrice des squamules d'*Elodea canadensis* à l'aide de quelques techniques permettant la détection des polysaccharides en microscopie électronique. *Journal Microscopie*, **8**, 82a.

Rougier, M. (1972). Etude cytochimique des squamules d'*Elodea canadensis*. Mise en évidence de leur sécrétion polysaccharidique et de leur activité phosphatasique acide. *Protoplasma*, **74**, 113–31.

Sawadis, T., Eleftheriou, E.P. & Tsekos, I. (1989). The floral nectaries of *Hibiscus rosa-sinensis*. III. A morphometric and ultrastructural approach. *Nordic Journal of Botany*, **9**, 63–71.

Scala, J., Schwab, D. & Simmons, E. (1968). The fine structure of the digestive gland of Venus's flytrap. *American Journal of Botany*, **55**, 649–57.

Schnepf, E. (1961*a*). Licht- und elektronenmikroskopische Beobachtungen an Insektivoren-Drüsen über die Sekretion des Fangschleimes. *Flora (Jena)*, **151**, 73–87.

Schnepf, E. (1961*b*). Quantitative Zusammenhänge zwischen der Sekretion des Fangschleimes und den Golgi-Strukturen bei *Drosophyllum lusitanicum*. *Zeitschrift für Naturforschung*, **16**, 605–10.

Schnepf, E. (1963). Zur Cytologie und Physiologie pflanzlicher Drüsen. 1. Teil. Über den Fangschleim der Insektivoren. *Flora (Jena)*, **153**, 1–22.

Schnepf, E. (1964). Zur Cytologie und Physiologie pflanzlicher Drüsen. 5. Teil. Elektronenmikroskopische Untersuchungen an Cyathialnektarien von *Euphorbia pulcherrima* in verschiedenen Funktionzustanden. *Protoplasma*, **58**, 193–219.

Schnepf, E. (1965*a*). Licht- und elektronenmikroskopische Beobachtungen an den Trichom-Hydatoden von *Cicer arietinum*. *Zeitschrift für Pflanzenphysiologie*, **53**, 245–54.

Schnepf, E. (1965*b*). Die Morphologie der Sekretion in pflanzlichen Drüsen. *Berichte der deutschen botanischen Gesellschaft*, **78**, 478–83.

Schnepf, E. (1965*c*). Morphologie der Duftölausscheidung bei *Typhonium divaricatum* (Araceae). *Planta*, **66**, 374–6.

Schnepf, E. (1968). Zur Feinstruktur der Schleimsezernierenden Drüsenhaare auf der Ochrea von *Rumex* and *Rheum*. *Planta*, **79**, 22–34.

Schnepf, E. (1969). Sekretion und Exkretion bei Pflanzen. *Protoplasmatologia*, **8**(8).

Schnepf, E. (1973). Sezernierende und exzernierende Zellen bei Pflanzen. In *Grundlagen der Cytologie*, ed. G.C. Hirsch, H. Ruska & P. Sitte, pp. 461–77. Stuttgart: Fischer Verlag.

Schnepf, E. (1974). Gland cells. In *Dynamic Aspects of Plant Ultrastructure*, ed. A.W. Robards, pp. 331–57. Maidenhead: McGraw Hill.

Schnepf, E. (1975). Special cytology: morphology and morphogenesis of cells of higher plants. In *Progress in Botany*, ed. H. Ellenberg, K. Esser, H. Merxmüller, E. Schnepf & H. Ziegler, pp. 30–6. Berlin: Springer Verlag.

Schnepf, E. (1976). Gland cells. In *Dynamic Aspects of Plant Ultrastructure*, ed. A.W. Robards, pp. 331–57. Maidenhead: McGraw-Hill.

Schnepf, E. & Benner, U. (1978). Die Morphologie der Nektarausscheidung bei Bromeliaceen. II. Experimentelle und qualitative Untersuchungen bei *Billberiga nutans*. *Biochemie und Physiologie der Pflanzen*, **173**, 23–36.

Schnepf, E. & Busch, J. (1976). Morphology and kinetics of slime secretion in glands of *Mimulus tilingii*. *Zeitschrift für Pflanzenphysiologie*, **79**, 62–71.

Schnepf, E. & Christ, P. (1980). Unusual transfer cells in the epithelium of the nectaries of *Asclepias curassavica* L. *Protoplasma*, **105**, 135–48.

Schnepf, E., Deichgräber, G. & Barthlott, W. (1983). On the fine structure of the liquid producing floral gland of the orchid, *Coryanthes speciosa*. *Nordic Journal of Botany*, **3**, 479–91.

Schnepf, E. & Pross, E. (1976). Differentiation and redifferentiation of a transfer cell: development of septal nectaries of *Aloe* and *Gasteria*. *Protoplasma*, **89**, 105–15.

Sedgley, M. & Blesing, M.A. (1983). Developmental anatomy of the avocado stigma papilla cells and their secretion. *Botanical Gazette*, **144**, 185–90.

Shimony, C. & Fahn, A. (1968). Light- and electron-microscopical studies on the structure of salt glands of *Tamarix aphylla* (L.). *Journal of the Linnean Society*, **60**, 283–8.

Shimony, C., Fahn, A. & Reinhold, L. (1973). Ultrastructure and ion gradients in the salt glands of *Avicennia marina* (Forsk.) Vierh. *New Phytologist*, **72**, 27–36.

Stern, W.L., Curry, K.J. & Pridgeon, A.M. (1987). Osmophores of *Stanhopea* (Orchidaceae). *American Journal of Botany*, **74**, 1323–31.

Terry, B.R. & Robards, A.W. (1987). Hydrodynamic radius alone governs the mobility of molecules through plasmodesmata. *Planta*, **171**, 145–57.

Thomson, W.W., Berry, W.L. & Liu, L.L. (1969). Localization and secretion of salt by the salt glands of *Tamarix aphylla*. *Proceedings of the National Academy of Sciences*, **63**, 310–17.

Thomson, W. W. & Liu, L. L. (1967). Ultrastructural features of the salt gland of *Tamarix aphylla* L. *Planta*, **73**, 201–20.

Trachtenberg, S. & Fahn, A. (1981). The mucilage cells of *Opuntia ficus-indica* (L.) Mill. Development, ultrastructure, and mucilage secretion. *Botanical Gazette*, **142**, 206–13.

Unzelman, J. M. & Healey, P. L. (1974). Development, structure and occurrence of secretory trichomes of *Pharbitis*. *Protoplasma*, **80**, 285–303.

Vassilyev, A. E. & Muravnik, L. E. (1988). The ultrastructure of the digestive glands in *Pinguicula vulgaris* (L.) (Lentibulariaceae) relative to their function. I. The changes during maturation. *Annals of Botany*, **62**, 329–41.

Vassilyev, A. E. & Stepanova, A. A. (1990). The ultrastructure of ion-secreting and non-secreting salt glands of *Limonium platyphyllum*. *Journal of Experimental Botany*, **41**, 41–6.

Vermeer, J. & Peterson, R. L. (1979). Glandular trichomes on the inflorescence of *Chrysanthemum morifolium* cv. Dramatic (Compositae). II. Ultrastructure and histochemistry. *Canadian Journal of Botany*, **57**, 714–29.

Wollenweber, E., Egger, K. & Schnepf, E. (1971). Flavenoid-Aglykone in *Alnus*-Knospen und die Feinstruktur der Drüsenzellen. *Biochemie und Physiologie der Pflanzen*, **162**, 193–202.

Wollenweber, E. & Schnepf, E. (1970). Vergleichende Untersuchungen über die flavonoiden Exkrete von 'Mehl'- und 'Öl'-Drüsen bei Primeln und die Feinstruktur der Drüsenzellen. *Zeitschrift für Pflanzenphysiologie*, **62**, 216–27.

Wrischer, M. (1962). Electronenmikroscopische Beobachtungen an extrafloralen Nektarien von *Vicia faba* L. *Acta Botanica Croatica*, **20/21**, 75–94.

Zamski, E., Shoham, G., Palevitch, D. & Levy, A. (1987). Ultrastructure of capsaicinoid-secreting cells in purgent and nonpurgent red pepper (*Capsicum annuum* L.) cultivars. *Botanical Gazette*, **148**, 1–6.

Ziegler, H. (1965). Die Physiologie pflanzlicher Drüsen. *Berichte der deutschen botanischen Gesellschaft*, **78**, 466–77.

Ziegler, H. & Lüttge, U. (1966). Die Salzdrüsen von *Limonium vulgare*. I. Die Feinstruktur. *Planta*, **70**, 193–206.

DAVID T. CLARKSON

New directions for membrane traffic

The chapters in this book serve to illustrate how radically the perception of biological membranes has changed in the last few decades. Membranes, when they were first considered by physical chemists and later by biophysicists, were seen largely as barriers to the free movement of solutes into cells and their compartments. The first electron microscopic studies of the unit membrane did little to change this view, although it was quickly appreciated that the specific characteristics of membranes must relate to structures or molecules which had not been visualised. Among the first to point out the dynamic nature of the plasma membrane (PM) of plant cells were cell biologists with an interest in plasmolysis. Stadelmann (1969) drew attention to the great rapidity with which the PM of onion epidermal cells and protoplasts could be disassembled and reassembled during osmotic excursions; in the present book such ideas are developed by Steponkus and by Oparka *et al.* Despite this evidence, plant scientists, especially transport physiologists, continued to regard such behaviour as an aberration created by the loss of turgor. Turgor, it seemed, was what distinguished higher plant cells from those of animals, and the PM was perceived as a rigid unyielding thing pressed firmly against the cell wall. This view took little account of what had already been established about the role of exocytotic vesicle traffic in wall building.

All of this contrasts with our current view in which membranes 'flow' between organelles in the form of vesicles. The membranes and their properties are now integrated into nearly every process in the life of cells, from early developmental events through to the final degradation of waste products by vesicular transport to lytic compartments. In this growth of understanding of the plant cell we have been led by the conclusions drawn from elegant experiments on animal cells and yeasts. It is clear that there are no fundamental differences in the way that cells of different types create vesicles, permit membrane fusion, secrete or store products. It is in matters of detail that characteristic behaviour has its origin. Homologies at the molecular level can be very useful clues for scientists working with less well characterised systems but they can sometimes be misleading, as Harris & Watson (this volume) point out. The peptides involved in protein-sorting in yeasts and animal cells are highly conserved and yet the actual mechanism of sorting differs markedly. Those working with plant cells in other areas have sometimes applied knowledge from animal cells rather too literally, e.g. in cell signalling.

The purpose of this final chapter is two-fold: first, to attempt to identify themes running through this book and, secondly, to indicate matters which the authors raised at an informal discussion at the end of the symposium. It is written from the perspective of a fairly conventional ion transport person, which at once explains its bias and naïvety.

Vesicle sorting and recognition

Molecular biology has provided a conceptual basis for understanding how the rather bewildering patterns of vesicle traffic are managed. Studies of protein targeting show that the insertion of proteins into the correct organelles or to the correct membrane involves specific amino acid sequences. Recognition is brought about by the matching or interaction of specific sequences, or motifs, on polypeptides. Thus, the assembly of coated pits around receptor molecules in the PM of animal cells may depend on the interaction of polypeptides of the clathrin coat with some motif on the cytoplasmic domains of membrane-bound receptors. From this we might assume that coated pits and vesicles which form elsewhere in the cell – e.g. on the *trans*-cisternae of the Golgi apparatus (GApp) – have different polypeptides which recognise different motifs. Clearly there is a pressing need to establish the molecular factors which determine where coated pits form. In discussion all contributors agreed that no-one had ever seen coated pits on the tonoplast membrane. This indicates that there is nothing on its surface which is recognised by the proteins of the clathrin coat.

Removal of clathrin coats from vesicles seems an essential step in the fusion of vesicles with larger organelles. Neither process occurs spontaneously. Uncoating requires the activity of a soluble ATPase and other ill-defined factors. Once uncoated, the docking of the vesicle with the partially coated reticulum or multi-vesicular bodies must again depend on the type of specific recognition mentioned above combined with other accessory factors. The scope for specificity in such processes can be seen in the example of *Saccharomyces cerevisiae* given by Harris & Watson (this volume). The secretory pathway for vesicular traffic from the endoplasmic reticulum (ER) through the GApp to ultimate destinations involves at least 50 different proteins. The step from ER to GApp alone involves ten SEC genes, for each of which mutants have been selected. Their existence has been demonstrated by a formidable combination of molecular genetics with cell biology and biochemistry.

Techniques have been developed for reconstruction of vesicular traffic between organelles *in vitro*. Gentle lysis of yeast protoplasts yields a heterogeneous mixture of organelles and microsomes which can be separated from cytosolic proteins and small molecules. The rate of transfer of a labelled precursor from one organelle to another can be studied after adding back cytoplasmic factors (for a review, see Gruenberg & Howell 1989). A further refinement of this type of procedure involves immobilising the membrane system which will accept vesicles, by attaching it to nitrocellulose filters and incubating it in solutions containing donor vesicles which are loaded with a labelled protein (Nowak *et al.* 1987). The mixture is incubated with or without ATP and various cytoplasmic factors. Recently there has been a report of the successful application of this method to membranes separated from leek seedlings, where ER was the donor membrane and the GApp the acceptor, bound to the nitrocellulose filter. In this study the authors were interested in the transfer of labelled lipids between the two systems (Moreau *et al.* 1990). Evidently, such re-constructions are going to play a very important role in exposing the finer points of vesicle traffic.

Endocytosis: routine and receptor mediated

Steer & O'Driscoll (this volume) have emphasised the necessity of recycling membrane which reaches the PM during the exocytosis of wall-building materials in growing cells. Endocytosis *via* coated vesicles (CVs) appears to have the capacity to handle the return flow of membranes, so that it has been estimated that whole PM can be turned over in 20–40 min in growing cells. Once a cell has completed its expansion, secondary wall development and maintenance requirements make much smaller demands on secretory processes and the abundance of coated pits and CVs decreases, the membrane turnover time increasing to 60–115 min (Emons & Traas 1986).

It has been known since at least 1954 (see Coleman *et al.* 1988) that unexpectedly large molecules and particles will enter plant cells and, as we can see in the chapter by Fowke *et al.*, such material may be firmly enclosed in CVs. It is probable that such entry is accidental and draws attention to the routine nature of endocytosis. Most cells can be found to contain CVs without being challenged by large molecules such as ferritin.

Returning to an earlier point, therefore, we must assume that clathrin-coat proteins are recognising and binding to something on the PM all the time. Clearly this makes a convincing proof of receptor-mediated endocytosis rather tricky, unless it is assumed that the routine endocytosis resulted from surface receptors being constantly triggered and recycled (there are precedents for this idea, see Goldstein *et al.* 1985). There is, however, one clear report of internalisation of a specific receptor in *intact* soybean cells. Horn *et al.* (1989) showed that fungal elicitor molecules from *Verticillium dahliae* became bound to specific sites on the surface of soybean PM and subsequently the sites (receptors) + fluorescently tagged elicitor were internalised by endocytosis. The authors did not verify that the internalised receptors were enclosed in CVs but it is likely that they were. Recently, the same authors (Horn *et al.* 1990) have shown that biotin, and fluorescently labelled proteins conjugated with biotin, are internalised rapidly by soybean cells. Biotin is a vitamin which has been shown in animal cells to have a well-characterised receptor which becomes internalised by endocytosis (Vesely *et al.* 1987). Quite apart from the usefulness of a system by which biotinylated macromolecules can be taken up by *intact* cells, the results of Horn *et al.* (1990) suggest that plant cells also have a receptor for this vitamin.

Some calculations suggest that the formation of intrusions of plasma membrane into the cytoplasm, against the direction of turgor pressure, is highly improbable, at least on the scale necessary to account for the fluxes of ions across plasma membranes (Cram 1980). These calculations were made in advance of knowledge about the pinching-off action during the assembly of the clathrin coat. Since this process is spontaneous and requires no energy input from ATP, we must assume that the free energy released by coat assembly is greater than the energy required to oppose the turgor forces. It is hard to see, except by theoretical consideration of molecular models, how the forces generated by coat assembly can be estimated. It may be, however, that turgor pressure is seriously over-estimated if the osmotic potential of the solutes in the apoplast is ignored. It is clear that certain solutes may become more concentrated in the apoplast than in the surroundings (see Grignon & Sentenac 1991).

Endocytosis: a contributor to solute fluxes into cells?

Even when they concede that endocytosis occurs, workers on ion transport are usually dismissive about the quantitative aspects of the process. Are they right? The following calculation of the volume of solution enclosed by CVs uses information, to be found elsewhere in this book, on the frequency of CVs; this has been combined with reasonable estimates

Table 1. *Calculation of possible Ca^{2+} inflow by endocytosis and comparison with observed fluxes of Ca^{2+} across plasma membranes*

Assumptions
(1) $[Ca^{2+}]$ free near membrane surface, 1 or 10 mol m^{-3} (10^{-18} or 10^{-17} mol μm^3)
(2) Diameter of CV, 0.05 μm; volume of CV lumen is 80% of total volume
(3) Rate of formation of CVs (see Steer & O'Driscoll, this volume):
2333s^{-1} (epidermal cell)

Calculate
(4) Volume of solution contained in CVs
0.21 $\mu m^3\ s^{-1}$ (pollen tube)
0.096 $\mu m^3\ s^{-1}$ (epidermal cell)
(5) Equivalent quantity of Ca^{2+} entering via CVs:
0.21 or 2.1×10^{-18} mol s^{-1} (pollen tube)
0.096 or 0.96×10^{-18} mol s^{-1} (epidermal cell)

Assume[a]
(6) Dimensions of pollen tube are $r = 5$ μm and $l = 1000$ μm giving volume of 78555 μm^3; surface area 0.314 mm^2
(7) Dimensions of epidermal cell are $r = 20$ μm and $l = 100$ μm giving volume of 12 568 μm^3; surface area 0.126 mm^2
(8) Cytosol occupies 10% of cell volume
(9) $[Ca^{2+}]$ free in cytoplasm is 10^{-4} mol m^{-3} (10^{-22} mol μm^3)

Calculate
(10) Free Ca^{2+} contents of cytoplasm:
0.78×10^{-18} mol (pollen tube) } Compare
0.13×10^{-18} mol (epidermal cell) } values in (5)
(11) Flux of Ca^{2+} delivered in CVs relative to PM surface area:
0.67 or 6.7×10^{-18} mol $mm^{-2}\ s^{-1}$ (pollen tube)[b]
0.76 or 7.6×10^{-18} mol $mm^{-2}\ s^{-1}$ (epidermal cell)[b]

Notes:
[a]l, length; *r*, radius.
[b]These values compare with steady-state efflux of 500×10^{-18} mol $mm^{-2}\ s^{-1}$ in mature cortical cells of onion (Macklon & Sim 1981) and 1600×10^{-18} mol $mm^{-2}\ s^{-1}$ in stomatal guard cells (MacRobbie 1989).

of periplasmic $[Ca^{2+}]$ and the results compared with Ca^{2+} fluxes across the plasma membrane (Table 1). Indeed, we must conclude that the Ca^{2+} inflow via CVs is small relative to measured fluxes into cells. It is, at most, 1.5% of that measured during steady-state Ca^{2+} efflux from a non-growing tissue (onion) and around 0.1% of that measured in metabolically active tissue. Only in one respect does the Ca^{2+} inflow via CVs seem to be significant. Clearly, the delivery each second would add very significantly to the cytosolic free Ca^{2+} activity; this is maintained at very low levels ($\sim 10^{-7}$ M) by a combination of sequestration and efflux pumping. But the vesicles release their contents into components of the endomembrane system, not into the cytoplasm directly. Calcium was chosen for the calculation because its concentration near to the external surface of the PM is likely to be far higher than that of other ions, because of electrostatic considerations and the activity of Ca^{2+} extrusion pumps at the PM. For other nutritionally important ions, therefore, whose fluxes are greater and whose periplasmic concentration is lower, endocytosis can be of no significance at all.

Vesicle traffic and central vacuole

Perceptions of the role of the vacuole influence the way one thinks about transport processes to or from it. It can be variously regarded as a large compartment in which undesirable things are dumped, worn-out things are broken down for recycling, useful things are stored or it can be seen as a well-managed reservoir of essential materials which flow in and out so as to maintain the ionic and metabolic equilibrium of the cytoplasm. The tonoplast is not subjected to the same physical constraints as the PM and yet, in mature cells, its area appears to remain relatively constant despite the fusion of membrane material from the vesicles which deliver material for processing or storage. It is only in rather special circumstances, such as the intense delivery of protein for storage, that the vacuole divides and the tonoplast area expands. The discussions at the symposium indicated much uncertainty about the occurrence of cycling of the tonoplast in the way so widely accepted for the PM. It was agreed, however, that no-one had ever seen a coated pit on the tonoplast in any cell.

The presence of a tonoplast-type proton pump (the V-type ATPase) in CVs has an important bearing on this discussion even though it explains nothing (Robinson *et al.*, this volume). Perhaps the first question is where CVs get their V-type ATPase. Certainly not from the tonoplast since coated pits do not seem to form. Certainly not from the PM which does not have a proton pump of this type. Perhaps then, it comes from the coated pits seen on the *trans*-cisternae of the GApp. If this is so, it suggests that this rather complex pump, consisting of up to nine subunits (Parry *et al.* 1989), is elaborated in the ER or GApp and could be delivered to the tonoplast in vesicles, coated or uncoated. The V-type ATPase is a most important piece of machinery which energises the tonoplast transport of essential materials, e.g. the H^+/Ca^{2+} antiport system which is crucial in maintenance of the cytosolic pCa. The continuous fusion of vesicles containing this ATPase could result in an over-provision of pumps. To maintain a correct provision, pumps would have to be degraded (which seems wasteful) or recycled (which has not been detected). In the discussion several contributors emphasised that membrane traffic was by no means the only mechanism by which proteins and lipids arrive in the appropriate membranes. Direct insertion of polypeptides via targeting sequences and the role of lipid transfer proteins are well understood. The turnover of membrane components can be controlled by intrinsic lipase and protease activities.

Since tonoplast vesicles can be well purified from plant cells it seems that there is an opportunity to compare rates of lipid and V-type ATPase turnover in cells which might have different rates of vesicular traffic between the tonoplast and the endomembrane system.

Exocytosis

The large quantities of polysaccharide material delivered to developing walls in uncoated vesicles create the need for membrane recycling, as discussed earlier. It seems, however, that in addition to this large capacity flow there is also a movement of CVs from the GApp to the plasma membrane. Earlier work by Griffing *et al.* (1986) showed that these contained little polysaccharide but much glucan synthase I, which is used for cellulose biosynthesis in the periplasm. It is rather hard to understand why this separate system is required. From the reasoning outlined above, it seems unlikely that these CVs have a V-type ATPase, since the fusion of the vesicle with the PM would mislocate this pump. In discussion it was claimed that CVs could be shown to have a vanadate-sensitive E_1E_2-type ATPase. Could it be that this pump is also assembled in the GApp? This point cannot be answered, but it should be remembered that the PM internalised from coated pits may include areas containing proton pumps. If there is cycling of the PM proton pumps it could be functional in replacing 'worn-out' mechanisms rather than being accidental.

The role of the antibiotic monensin in blocking secretion from the GApp cisternae has been touched on in this book (Jones *et al.*). This ionophore brings about an exchange of protons for sodium ions and will collapse the ΔpH across membranes created by the proton pump. Its effects on secretion in animal cells have been studied extensively (Farquhar 1985) and suggest that the medial cisternae of the GApp normally have an acidic interior, and are presumably equipped with a proton pump. The processing and sorting of proteins in the GApp is blocked when the ΔpH across the cisternae membranes is collapsed by monensin. Another ionophore, nigericin, which brings about an electroneutral H^+/K^+ exchange, also disrupts secretion but is less effective than monensin in animal cells. One characteristic of monensin-treated secretory cells is the accumulation of many vesicles in the cytoplasm in which incompletely processed proteins may be found. Why do not these vesicles fuse with the PM or tonoplast in the usual way? Their accumulation suggests that their surface properties may be lacking the factors essential for fusion; these may relate to membrane polypeptides, surface charge and calcium ion concentration.

Conclusions

Cells which are growing rapidly or those which are secreting or storing copious amounts of protein are the most suitable model systems in which to study membrane traffic. If the model system is one in which mutants can be readily produced and characterised, much insight can be gained into the factors which control this traffic. It is reasonable to ask, however, whether the recycling of membranes occurs in mature cells whose metabolic activity proceeds at a more measured pace. In such cells the activity of membrane proteins continues, unwanted substances may enter the cell through the plasma membrane, enzymes wear out and need to be replaced, and proteolytic and hydrolytic activities remain centred in vacuoles and lysosomes. We need to establish whether the processes described in this book really operate in such cells.

Perhaps a final word should be left to a remark which I noted down in the discussion. It relates to the earlier concern that plant scientists may be overanxious to apply animal models to their own cells. A contributor pointed out 'Some roots contain haemoglobin, but plants don't have hearts.'

References

Coleman, J., Evans, D. & Hawes, C. (1988). Plant coated vesicles. *Plant, Cell and Environment*, **11**, 669–84.

Cram, W.J. (1980). Pinocytosis in plants. *New Phytologist*, **84**, 1–17.

Emons, A.M.C. & Traas, J.A. (1986). Coated pits and coated vesicles on the plasma membrane of plant cells. *European Journal of Cell Biology*, **41**, 57–64.

Farquhar, M.G. (1985). Progress in unravelling pathways of Golgi traffic. *Annual Review of Cell Biology*, **1**, 447–88.

Goldstein, J.L., Brown, M.S., Anderson, R.G.W., Russell, D.W. & Schneider, W.J. (1985). Receptor mediated endocytosis: concepts emerging from the LDL receptor system. *Annual Review of Cell Biology*, **1**, 1–39.

Griffing, L. R., Mersey, B. G. & Fowke, L. C. (1986). Cell fractionation analysis of glucan synthase I and II distribution and polysaccharide secretion in soybean protoplasts. Evidence for the involvement of coated vesicles in wall biogenesis. *Planta*, **167**, 175–82.

Grignon, C. & Sentenac, H. (1991). Properties of the apoplast. *Annual Review of Plant Physiology and Plant Molecular Biology*, **42** (in press).

Gruenberg, J. & Howell, K. E. (1989). Membrane traffic in endocytosis: insights from cell-free assays. *Annual Review of Cell Biology*, **5**, 453–81.

Horn, M. A., Heinstein, P. F. & Low, P. S. (1989). Receptor-mediated endocytosis in plant cells. *Plant Cell*, **1**, 1003–9.

Horn, M. A., Heinstein, P. F. & Low, P. S. (1990). Biotin-mediated delivery of exogenous macromolecules into soybean cells. *Plant Physiology*, **93**, 1492–6.

Macklon, A. E. S. & Sim, A. (1981). Cortical cell fluxes and transport to the stele in root segments of *Allium cepa*. IV. Calcium as affected by its external concentration. *Planta*, **152**, 381–7.

MacRobbie, E. A. C. (1989). Calcium efflux at the plasmalemma of isolated guard cells of *Commelina communis*. Effects of abscisic acid. *Planta*, **178**, 231–41.

Moreau, P., Bertho, P. & Cassagne, C. (1990). Involvement of the intracellular lipid traffic during plasma membrane biogenesis in higher plants. In *Dynamics and Biogenesis of Membranes*, ed. J. A. F. Op den Kamp, pp. 77–83. NATO ASI Series Vol. H40. Berlin: Springer-Verlag.

Nowack, D. D., Morré, D. M., Paulik, M., Keenan, T. W. & Morré, D. J. (1987). Intracellular membrane flow: reconstitution of transition vesicle formation and fusion in a cell free system. *Proceedings of the National Academy of Sciences, USA*, **84**, 6098–102.

Parry, R. V., Turner, J. C. & Rea, P. A. (1989). High purity preparations of higher plant vacuolar H^+-ATPase reveal additional subunits. Revised subunit composition. *Journal of Biological Chemistry*, **264**, 20 025–32.

Stadelmann, E. J. (1969). Permeability of the plant cell. *Annual Review of Plant Physiology*, **20**, 585–606.

Vesely, D. L., Kemp, S. P. & Elders, M. J (1987). Isolation of a biotin receptor from hepatic plasma membranes. *Biochemical and Biophysical Research Communications*, **143**, 913–16.

Index